walkernaths 1.2

Mathematical Methods

Charlotte Walker and Victoria Walker

Walker Maths: 1.2 Mathematical Methods
1st Edition
Charlotte Walker
Victoria Walker

Cover design: Cheryl Smith, Macarn Design
Text design: Cheryl Smith, Macarn Design
Production controller: Siew Han Ong

Any URLs contained in this publication were checked for currency during the production process. Note, however, that the publisher cannot vouch for the ongoing currency of URLs.

Acknowledgements
The authors and publisher wish to thank Beth Hunter from Poi Princess for the cover image. All other images are courtesy of Shutterstock and iStock.

We also wish to acknowledge the trusted kindred spirits throughout the country for your help in preparing this title.

© 2023 Cengage Learning Australia Pty Limited

For product information and technology assistance,
    in Australia call **1300 790 853**;
    in New Zealand call **0800 449 725**

For permission to use material from this text or product, please email
aust.permissions@cengage.com

National Library of New Zealand Cataloguing-in-Publication Data
A catalogue record for this book is available from the National Library of New Zealand.

978 01 7047746 8

Cengage Learning Australia
Level 5, 80 Dorcas Street
Southbank VIC 3006 Australia

Printed in China by 1010 Printing International Limited.
3 4 5 6 7 27 26 25 24

This icon appears throughout the workbook. It indicates that there is a worksheet available which has been collated from the original Level 1 WalkerMaths series. The worksheets can be accessed via the Teacher Resource for this workbook (available for purchase from nz.sales@cengage.com). These worksheets are an additional resource which can be used to support your students throughout the teaching of this standard.

CONTENTS

Number

Fundamentals

Rounding — decimal places

- When asked to round to 2 dp (two decimal places), this means there should be **exactly** two digits after the decimal point.

Locate the digit to the **right** of the last required decimal place.

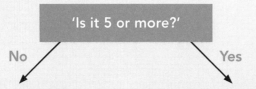

'Is it 5 or more?'

No   Yes

Leave the previous digit alone   **Increase the previous digit by one**

Example 1: Round 3.1624 to 2 dp.

The 2 is **not** 5 or more
$\Rightarrow$ leave the 6 alone

$\therefore$ 3.1624 = 3.16 (2 dp)

Example 2: Round 19.2463 to 2 dp.

The 6 **is** 5 or more
$\Rightarrow$ increase the 4 to a 5

$\therefore$ 19.2463 = 19.25 (2 dp)

Rounding — significant figures

How to round to significant figures

- When asked to round to 4 sf (four significant figures), this means there should be exactly four significant figures in the answer.
- The process is almost the same as rounding to decimal places.
- **Always** check that the rounded number is a similar size to the original number.

Locate the digit to the **right** of the last required significant figure.

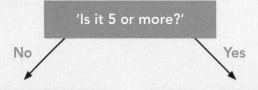

'Is it 5 or more?'

No   Yes

Leave the previous digit alone   **Increase the previous digit by one**

Example 1: Round 2.09314 to 4 sf.

The 1 is **not** 5 or more
$\Rightarrow$ leave the 3 alone

$\therefore$ 2.09314 = 2.093 (4 sf)

Example 2: Round 24.9274 to 4 sf.

The 7 **is** 5 or more
$\Rightarrow$ increase the 2 to a 3

$\therefore$ 24.9274 = 24.93 (4 sf)

Rates and proportions

- A rate is a relationship between two quantities.
- Rates are often written using the word '**per**'.
- '**Per**' means 'for each' and is another way of saying **divide**.
- Examples: the amount ($) paid per hour for gardening;
  the speed at which a person walks in kilometres per hour.
- When we use two rates in a calculation, we assume that the relationship between the quantities in each remains constant. We assume that they are **in proportion**.

Example: 8 kg of kūmara cost $23.92. Calculate the cost of 3 kg of kūmara.

Method 1: Find the value of one unit and multiply this by the number of units you need.

8 kg cost **$23.92**, so 1 kg costs $23.92 ÷ **8** = **$2.99**

3 kg cost **3** x $2.99 = **$8.97**

> One unit (1 kg) costs $2.99.

Method 2: Use fractions.

> Because you are trying to find a price, put the **price** on the **right**.

8 kg cost **$23.92**

copy ↓

3 kg cost **$23.92** x $\frac{3}{8}$ = **$8.97**

> Multiply by either $\frac{3}{8}$ or $\frac{8}{3}$. Use $\frac{3}{8}$ here because 3 kg will cost **less**.

Answer the following questions.

1 If 4 kg of oranges cost $23.96, how much would 3 kg of oranges cost?

_____

_____

2 If 4 kg of mince cost $47.80, calculate the cost of 7 kg of mince.

_____

_____

3 Last year, Mere made 13 jars of jam from 3.25 kg of plums. How many jars can she expect to fill with 4.5 kg of plums?

_____

_____

4 If 210 million posts are created on a social media app every day, how many are posted each hour?

_____

_____

5 Hair grows at 3.5 mm every 10 days. How much does hair grow in a year (365 days)? Write your answer to the nearest millimetre.

_____

_____

6 One New Zealand dollar buys $US0.63. If Angus brings $US567.00 into New Zealand, how many New Zealand dollars will he get when he changes it?

_____

_____

Inverse proportions

- In the previous section, as one quantity got bigger, the other also got bigger, e.g. as the mass of kūmara increased, the cost also increased.
- In some relationships, as one quantity gets bigger, the other **becomes smaller**, e.g. the faster you travel to school, the shorter the time it takes you.

Example: Matilda bikes home from town. She can travel at 21 kph and it takes her 13 minutes. How long would it take her to walk home from town at 5 kph?

Method 1:   Find the value of one unit and multiply this by the number of units you need.

At **21** kph she takes **13** min, so at **1** kph it would take her **21** x 13 = **273** min

At **5** kph she will take 273 ÷ **5** = **54.6 min**

> Notice that you need to **multiply** because at 1 kph it will take her **longer**.

Method 2:   Use fractions.

At $\boxed{21}$ kph she takes **13 min**

copy ↓

At $\boxed{5}$ kph she takes $13 \times \dfrac{21}{5}$ = **54.6 min**

> Multiply by either $\dfrac{5}{21}$ or $\dfrac{21}{5}$. Use $\dfrac{21}{5}$ here because it will take her **longer** to walk.

Answer the following questions.

1   It took 11 minutes to fill the bath using just the cold tap. If both taps are used and they have similar flow rates, how long will it take to fill the bath?

2   A snail moving at 1 m per hour took 12 minutes to climb a wall. A beetle took 1.5 minutes to climb the same wall. How fast was the beetle moving?

3   On Monday, Hemi jogged home from school at 6 kph and it took him 40 min. On Tuesday he walked at 5 kph. How long did it take him?

4   It took a team of six painters $10\frac{1}{2}$ days to paint an office block. How long should it take a team of 7 painters to do the same job?

5   It took three and a half hours for 7 parents to make 480 lamingtons. If only 5 parents were doing the same job, how long (in hours and minutes) would it take them?

6   A spa pool takes $17\frac{1}{2}$ minutes to fill at a rate of 120 L per minute. How many minutes would it take to fill at a rate of 150 L per minute?

Assumptions and limitations

- When answering questions, we should discuss any **assumptions** we have made: we often assume certain conditions apply.

Examples: From the previous page:

2 We assumed that the snail and the beetle took the same route up the wall and we assumed that both climbed the wall in similar weather conditions.

3 We assumed that conditions were similar and there wasn't, for instance, a strong wind on one of the days. We assumed that his backpack was about the same mass on each day.

4 We assumed that all the painters worked at the same rate.

5 We assumed that all the parents made lamingtons at the same rate.

- We also need to discuss any **limitations** to our answers: when or to whom would our answer not apply.

Examples: From the previous page:

2 Our answer would apply only to snails and beetles of the same size and species.

3 Our answer would apply only to Hemi and this particular route, not to other routes which might require, for instance, waiting for light changes.

4 Our answer would apply only to this paint job on this building. Other paint jobs may involve different painting skills, such as painting around windows, using several colours, etc.

5 Our answer would apply only to making of lamingtons of the same size and shape.

Often, assumptions and limitations can overlap, so don't worry too much about distinguishing between them.

Answer the following questions.

1 a A recipe requires 3 cups of flour for 15 cookies. How many cookies could be made with 7 cups of flour?

_____

b Discuss any assumptions and limitations of your answer.

_____

_____

_____

2 a A flea can jump a distance of 200 times its body length. If Gia could jump the equivalent of a flea and she is 165 cm long, how far could she jump?

_____

b Discuss any assumptions and limitations of your answer.

_____

_____

_____

3 **a** Hemi can type 125 words in 5 minutes. How long would it take him to type a 1000-word document?

_____

b Discuss any assumptions and limitations of your answer.

_____

_____

_____

4 **a** Dirk is going to England. One New Zealand dollar buys £0.52 (pounds sterling). He needs £500.00 for when he first arrives. How much will this cost him in New Zealand dollars?

_____

b Discuss any assumptions and limitations of your answer.

_____

_____

_____

5 **a** There is an average of 126,000 spiders per hectare (10 000 m^2) in green areas. How many spiders could you expect to have on a rugby field (7000 m^2)?

_____

b Discuss any assumptions and limitations of your answer.

_____

_____

_____

6 **a** Albatross can circle the globe (40 000 km) in 46 days. How long would it take one to fly the length of New Zealand (2090 km)? Write your answer to the nearest hour.

_____

b Discuss any assumptions and limitations of your answer.

_____

_____

_____

Ratios

- Ratios show how an amount is split into several shares, often different sizes.
- Both parts of a ratio must use the **same units**.
- Ratios generally use whole numbers, not decimals or fractions, and are written in their simplest form.
- A **colon** (:) is used to separate the two numbers, e.g. 2:3 means 'two parts to three parts'.

Using ratios where the total is given

Example: $279 needs to be shared between two people in the ratio 4:5.

When there is a quantity that needs to be split using a ratio, follow these steps.

Step 1: Find the total number of parts by adding the numbers in each share: $4 + 5 = 9$

Step 2: Find the value of one part by dividing the total by the number of parts: $\$279 \div 9 = \$31$

Step 3: Multiply the value of one part by each part of the ratio: $4 \times \$31 : 5 \times \$31$

So the ratio of money is $\$124 : \$155$

So one person gets $124 and the other gets $155.

$279 split into 9

| $31 | $31 | $31 | $31 | $31 | $31 | $31 | $31 | $31 |

4 parts = $124   **5 parts = $155**

Another way to think of this is $\frac{4}{9}$ of $279 and $\frac{5}{9}$ of $279.

Note: While the ratio doesn't have units or decimals, your answer could have these.

Share the quantities in the given ratios.

1   16 kg in the ratio 5:3

2 $90 in the ratio 11:4

_____

_____

3 $108 needs to be split between Ali and Fred in the ratio of 4:5. How much should each person receive?

_____

_____

4 In New Zealand, the ratio of left-handed to right-handed people is 1:9. In a school with a roll of 1437 students, how many would you expect to be left handed?

_____

_____

5 In a beehive, the ratio of female worker bees to male drone bees is 110:1. If a hive contains 59 940 bees (not counting the queen), how many of these will be female worker bees?

_____

_____

6 In New Zealand, the ratio of pet owners to those that don't own a pet is 16:9. In 2023, the population was 5.2 million. How many pet owners were there?

_____

_____

7 Jeremy, Julia and Jai got paid $260 for some gardening. Jeremy worked for 3 hours, Julia for 4 hours and Jai for 1 hour. How much money should each of them receive?

_____

_____

8 Amanda is helping to make up food parcels for four families. The families consist of one with six people, two with four people, and one with two people. She has 12.8 kg of potatoes to share between them. What mass of potatoes should each family get?

_____

_____

9 Mila paid $3.00 towards a packet of 33 pineapple lumps, and Matiu paid the remaining $3.60. If the pineapple lumps are shared fairly between them, how many should each get?

_____

_____

10 Four families combined to buy a 10 kg bag of rice. One family paid $13.50, the second paid $18.00, a third paid $9.00, and the fourth paid $4.50. If the rice is divided fairly, how much rice should each family get?

_____

_____

     ISBN: 9780170477468

Ratio calculations where one part is given

Examples:

1 Finding a part

Two brothers were given surfboards. The ratio of the length of the younger brother's surfboard to the older brother's surfboard was 7:9. If the younger brother's board is 2.1 m long, how long is the older brother's board?

2.1 m

Step 1:	Calculate the value of each part:	2.1 ÷ 7 = 0.3
Step 2:	Multiply the value of each part (0.3) by the number of parts (9):	0.3 x 9 = 2.7

So the older brother's board was 2.7 m long.

2 Finding a total

Alia and Nikora pooled their money to buy an $8 bag of lollies. Alia had $5 and Nikora had $3. If they shared them fairly and Nikora got 42 lollies, how many lollies were in the bag?

Step 1:	Find the total number of parts by adding the numbers in each share (5 and 3):	5 + 3 = 8
Step 2:	Calculate the value of each part by dividing the number given (42) by the number of parts it represents (3):	42 ÷ 3 = 14
Step 3:	Multiply the value of each part (14) by the total number of parts (8):	14 x 8 = 112

So there were 112 lollies in the bag.

Answer the following questions.

1   The ratio of the length of a boat's anchor rope to the depth of water should be 3:1 so that a boat is held in place. A skipper wants to anchor where the water is 16 m deep. How much anchor rope should he let out?

_____

_____

2   The ratio of the leg span of the female katipō spider to that of the male is 6:1. Mature females have a leg span of 32 mm. What is the leg span of a male katipō spider?

_____

_____

3   The ratio of the number of species of spiders to the number of species of mammals is 25:3. If there are 5400 species of mammals, how many species of spiders are there?

_____

_____

4   The ratio of Australians that have at least one pet to those that don't own a pet is 7:3. If 18.2 million Australians have pets, how many Australians don't have pets?

_____

_____

5   The ratio of the height to the width of the Argentinian flag is 5:8. If an Argentinian flag is 104 cm high, how wide is it?

_____

_____

6   When making pastry, the ratios (by mass) of flour to butter to water are 3:2:1. Morgan has only 172 g butter and he wants to make as much pastry as he can. How much flour and water should he use?

_____

_____

7   Emily, Eru and Ernie were paid to paint the garage. Emily spent 8 hours on the job, Eru spent 5 hours and Ernie spent just 2 hours painting. If Eru was paid $90, how much were they paid altogether for the job?

_____

_____

8   The steepest street in the world is Baldwin Street in Dunedin. For its steepest section, the ratio of the height it rises to the horizontal distance it covers is 1:1.34. The street rises 47 m. What horizontal distance does it cover?

_____

_____

   ISBN: 9780170477468

Scale diagrams

- The scale tells us the ratio of a length in the picture or map compared with the corresponding length of the real object or the distance.
- Notice that these two lengths **must be in the same units**.

There are two types of scale factor.

> The numbers are separated by a colon.

1 The first number is smaller (usually 1) **:** the second number is bigger.
This means the diagram is smaller than the real object.

Example: Scale of 1:4 means 1 cm on the picture = 4 cm on the object.

Picture	**Real paperclip**

1 cm

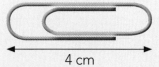

4 cm

2 The first number is bigger **:** the second number is smaller (usually 1).
This means the diagram is bigger than the real object.

Example: Scale of 4:1 means 4 cm on the picture = 1 cm on the object.

Picture	**Real pumpkin seed**

4 cm

1 cm

Examples:

1 The diagram shows triceratops dinosaur whose actual length is 8.1 m.

4 cm

9 cm

a Calculate the scale.

9 cm on diagram = 8.1 m in real life

$$1 \text{ cm on diagram} = \frac{8.1 \text{ m}}{9 \text{ cm}} = \frac{810 \text{ cm}}{9 \text{ cm}} = 90 \text{ cm in real life.}$$

So the scale is 1:90.

> The units **must** be the same.

b Calculate the actual height of the dinosaur.

9 cm on diagram = 8.1 m in real life

Triceratops height = 4 cm on diagram = $8.1 \times \dfrac{4}{9}$ = 3.6 m in real life.

2 This illustration of a Chevrolet Impala coupe is drawn to a scale of 1:50.
This means that 1 cm on the illustration = 50 cm on the car.

Measure the length of the car in the diagram and calculate the length of the real car.

1 cm on diagram = 50 cm on the car

Car length = 11.2 cm on the diagram = 50 x 11.2

= 560 cm

= 5.6 m

3 This diagram of a housefly is drawn to a scale of 5:1.
This means that 5 cm on the diagram = 1 cm on the fly.

Measure the length of the body (including the head) on the
diagram and calculate the length of the body of a housefly.

5 cm on diagram = 1 cm on the fly

Body length = 3 cm on diagram = $1 \times \dfrac{3}{5}$

= 0.6 cm

= 6 mm

Answer the following questions.

1 An image of an object is 40 mm wide.
The actual object is 800 mm wide.
Calculate the scale.

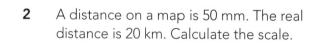

2 A distance on a map is 50 mm. The real
distance is 20 km. Calculate the scale.

3 An image of an object is 80 cm wide. The
actual object is 2 cm wide. Calculate the
scale.

4 An image of an object is 6 cm wide. The
actual object is 3 mm wide. Calculate the
scale.

 ISBN: 9780170477468

5 This motorbike was designed by New Zealander John Britten. The scale in this photograph is 1:40, and the length of the bike image is 5 cm.
How long is the actual bike?

_____

_____

6 Hamish has done a drawing of a biplane using a scale of 1:125. Measure the length of the biplane in the diagram and use it to calculate the length of the real biplane.

_____

_____

7 The scale in this photograph is 1:5. Measure the length of this image and calculate the actual length of this papa hou (treasure box).

_____

_____

8 The scale for this photograph of a mosquito is 5:1. Measure the total length of this mosquito, including its legs, and use it to calculate the length of a real mosquito.

_____

_____

9 This is a photograph of a single peppercorn. The scale is 15:1. Measure its maximum diameter, and use this to calculate the diameter of the actual peppercorn.

_____

_____

_____

_____

10 The scale for this map is 1:200 000. Measure the length of Lake Rotoroa on the map and use your measurement to calculate the actual length of Lake Rotoroa.

_____

_____

_____

_____

_____

11 The scale for this map is 1:500 000. Measure the length of Rangitoto ki te Tonga/D'Urville Island on the map and use your measurement to calculate its actual length.

_____

_____

_____

_____

ISBN: 9780170477468

12 This is a design for a tiny house. The scale is 1:50.

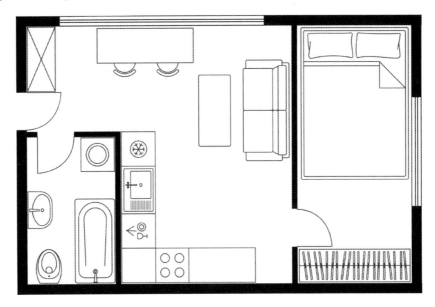

a Calculate the total length of the tiny home.

_____

b Will a bed that is 165 cm wide by 203 cm long fit in the bedroom? Explain your answer.

_____

_____

13 The diagram shows the plan for a garden.

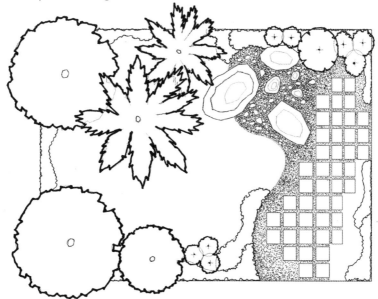

a The two large trees on the left have diameters of 3 m. Calculate the scale for this plan.

_____

b Use the scale to calculate the total length of the garden (not counting the overhang of the trees).

_____

Powers

Fractional powers and roots

What are roots?

Finding the **square** root of a number is the same as asking what **two** identical numbers multiply

to produce that number, e.g. $\sqrt{9} = \sqrt{? \times ?}$

$$= \sqrt{3 \times 3}$$

$$= 3$$

Finding the **cube** root of a number is the same as asking what **three** identical numbers multiply

to produce that number, e.g. $\sqrt[3]{27} = \sqrt{? \times ? \times ?}$

$$= \sqrt{3 \times 3 \times 3}$$

$$= 3$$

Roots can also be written as **fractional powers**.

Examples:

These are different ways of writing the same thing.

1 $25^{\frac{1}{2}} = \sqrt[2]{25}$

$$= \sqrt{25}$$

$$= 5$$

2 $64^{\frac{1}{3}} = \sqrt[3]{64}$

$$= 4$$

You can find the value of a root or fractional power using your calculator.

Cube roots: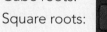
Square roots: Any other roots:

Rewrite these numbers using root signs, and find their values.

1 $16^{\frac{1}{2}} = $ _____

$= $ _____

2 $49^{\frac{1}{2}} = $ _____

$= $ _____

3 $0.04^{\frac{1}{2}} = $ _____

$= $ _____

4 $125^{\frac{1}{3}} = $ _____

$= $ _____

5 $0.027^{\frac{1}{3}} = $ _____

$= $ _____

6 $1728^{\frac{1}{3}} = $ _____

$= $ _____

7 $32^{\frac{1}{5}} = $ _____

$= $ _____

8 $0.0256^{\frac{1}{4}} = $ _____

$= $ _____

 ISBN: 9780170477468

Negative powers

- These are the reciprocal of the positive power.
- To find the reciprocal, you turn the fraction upside down.

Examples: **1** $4^{-2} = \left(\frac{1}{4}\right)^2 = \frac{1^2}{4^2} = \frac{1}{16}$ **2** $(-2)^{-3} = \left(\frac{1}{(-2)}\right)^3 = \frac{1^3}{(-2)^3} = -\frac{1}{8}$

3 $\left(\frac{2}{5}\right)^{-3} = \left(\frac{5}{2}\right)^3 = \frac{5^3}{2^3} = \frac{125}{8}$

6

Rewrite, then calculate the values of these powers.

1 $2^{-1} =$ _____

        $=$ _____

2 $10^{-2} =$ _____

        $=$ _____

3 $(0.2)^{-3} =$ _____

        $=$ _____

4 $\left(\frac{3}{4}\right)^{-2} =$ _____

        $=$ _____

Combining negative and fractional powers

Examples:

1 $27^{-\frac{1}{3}} = \left(\frac{1}{27}\right)^{\frac{1}{3}} = \frac{1^{\frac{1}{3}}}{27^{\frac{1}{3}}} = \frac{1}{3}$ **2** $\left(\frac{16}{25}\right)^{-\frac{1}{2}} = \left(\frac{25}{16}\right)^{\frac{1}{2}} = \sqrt{\frac{25}{16}} = \frac{\sqrt{25}}{\sqrt{16}} = \pm\frac{5}{4}$

Rewrite, then calculate the values of these powers. Leave your answer as a fraction.

1 $16^{-\frac{1}{2}} =$ _____

        $=$ _____

2 $243^{-\frac{1}{5}} =$ _____

        $=$ _____

3 $(1\,000\,000)^{-\frac{1}{3}} =$ _____

        $=$ _____

4 $\left(\frac{1}{121}\right)^{-\frac{1}{2}} =$ _____

        $=$ _____

5 $\left(\frac{4}{9}\right)^{-\frac{1}{2}} =$ _____

        $=$ _____

6 $\left(\frac{8}{125}\right)^{-\frac{1}{3}} =$ _____

        $=$ _____

Use your calculator to find the value of the following. Round your answers to 3 sf.

7 $8^{\frac{2}{3}} =$ _____

8 $10\,000^{\frac{3}{4}} =$ _____

9 $4^{1.2} =$ _____

10 $7^{0.3} =$ _____

Percentages

Using percentages

Examples:

1 Finding a percentage of a quantity

Find 4% of 450.

Remember, 'of' means multiply.

You could: convert the percentage to a decimal:

4% of $450 = 0.04 \times 450$

$= 18$

Or: use the % function on your calculator:

$4\% \times 450 = 18$

Not all calculators are the same, so you will need to experiment until you find how yours works.

2 Increasing by a percentage

You could use the % button on your calculator.

Increase $80 by 12%.

This means that we need 100% plus 12%.

Increased amount $= \$80 + (12\% \text{ of } 80)$

$= \$80 + 9.6$

$= \$89.60$

or   Increased amount $= \$80 \times 112\%$

$= \$80 \times 1.12$

$= \$89.60$

3 Decreasing by a percentage

Note: With money, always round to 2 dp.

Decrease 160 by 24%.

This means that we need 100% minus 24%, or 76%.

Decreased amount $= 160 - (24\% \text{ of } 160)$

$= 160 - 38.4$

$= 121.6$

or   Decreased amount $= 160 \times 76\%$

$= 160 \times 0.76$

$= 121.6$

4 Calculating a percentage

A watermelon of mass 2.5 kg contains 2.275 kg of water. What percentage of the melon is water?

$$\text{Percentage} = \frac{2.275}{2.5} \times 100$$

$$= 91\%$$

Calculate these.

1   35% of 65 km = _____

2   7% of $1560 = _____

3   75% of 950 mL = _____

4   2% of $9905 = _____

5   Increase 50 km by 24% = _____

6   Decrease $90 by 25% = _____

7   Increase 75 kg by 15% = _____

8   Increase 640 g by 8% = _____

9   Decrease $900 by 60% = _____

10   Decrease 250 mL by 2% = _____

Answer the following questions.

11   A hockey team played 28 games in a season. It won 75% of them. How many games did the team win?

_____ games

12   A school has 944 students on its roll and 12.5% of them are left handed. How many are left handed?

_____ students

13   Luka currently earns a wage of $24 an hour. He has just been given a 5% pay rise. What is his new hourly rate?

$_____

14   The value of a new car decreases by 10% as soon as it is bought and driven off the car yard. Tamati just bought a car for $15 500. What is its value after he has driven it away?

$_____

15   Auntie Miriam revised her will and left 35% of her estate to charity. The remainder is shared equally between each of her four sons. If her estate is worth $46 000, how much does each son inherit?

$_____

16   A shop advertisement states there is 30% off everything. What is the sale price of a jacket that normally costs $499.90?

$_____

17   It is estimated that there are 8.2 million species of organisms on Earth, of which 6.96 million have not yet been discovered. What percentage of species has been discovered?

_____%

18   Great white sharks can detect one drop of blood in 100 L of water. There are 10 drops in one millilitre. Calculate the percentage of blood in the water.

_____%

Calculating percentage changes

Examples:

1 Percentage increase

The value of a sculpture increased from $480 to $648. Calculate the percentage increase in its value.

Step 1: **Subtract** the amounts: $648 − $480 = $168

Step 2: **Divide** the difference by the **original** quantity, then multiply by 100. $\frac{168}{480} \times 100 = 35\%$

2 Percentage decrease

The original price of some jeans was $98, and the sale price is $78.40. Calculate the percentage reduction in its price.

Step 1: **Subtract** the amounts: $98 − $78.40 = $19.60

Step 2: **Divide** the difference by the **original** quantity, then multiply by 100. $\frac{19.6}{98} \times 100 = 20\%$

Calculate the percentage changes below.

1   From 35 to 56 = +/– _____%

2   From 230 to 138 = +/– _____%

3   From $380 to $285 = +/– _____%

4   From 190 g to 218.5 g = +/– _____%

5   From $2150 to $2537 = +/– _____%

6   From 97.2 kg to 37.9 kg = +/– _____%

Answer the following questions.

7   A pair of jeans was $72, but is now $54. Calculate the percentage reduction.

_____

8   The population of a town increased from 4186 to 6342. Calculate the percentage increase in the population.

_____

9   Lou bought a phone for $990, but a couple of months later sold it for $560. Calculate her percentage loss on the phone.

_____

10   Ari's rent increased from $260 a week to $275 a week. Calculate the percentage increase.

_____

GST

- **GST** stands for **G**oods and **S**ervices **T**ax.
- It is added to everything you buy and it goes to the Government to fund the running of the country.
- The standard current GST rate on all products and services is **15%**.
- Don't forget to add units and round sensibly.

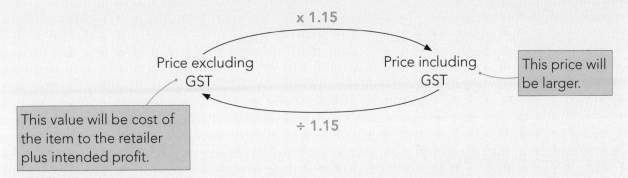

Price excluding GST — x 1.15 → Price including GST

This price will be larger.

This value will be cost of the item to the retailer plus intended profit.

÷ 1.15

Examples:

1 Finding a GST inclusive price

The pre-GST price of a phone is $1738.00. Calculate the retail price of the phone.

GST inclusive price = $1738 x 1.15

= $1998.70

This is the price including GST.

2 Finding the amount of GST paid on an item

A bed costs $1490 including GST. How much GST was paid?

GST exclusive price = $1490 ÷ 1.15

= $1295.65 (2 dp)

This is the price without GST.

∴ GST = $1490 – $1295.65

= $194.35

Add GST to these amounts to find their selling price.

1 $58 _____

2 $783 _____

3 $10 008 _____

4 $516 000 _____

These prices include GST. Calculate their GST exclusive prices.

5 $19.60 _____

6 $16 050 _____

7 $419.00 _____

8 $83 000 _____

These prices include GST. Calculate the amount of GST that is included in each.

9 $276.00 _____

10 $8.74 _____

11 $1159.20 _____

12 $49 220 _____

Answer the following questions.

13 A TV is advertised at $674.99. Calculate its pre-GST value. _____

14 How much GST did Laura pay on her pie that cost her $5.80? _____

15 Beatrice bought a new bike that cost her $1060. How much GST was included in this price?

_____

16 An electrician charges Greg $450 plus GST. How much does Greg have to pay the electrician?

_____

17 Chloe buys a wheelbarrow for $259 including GST. How much of this price is GST?

_____

18 A haircut cost Fiona $26. What was the pre-GST price? _____

19 A hardware store has 4 L of paint priced at $132 excluding GST. How much will the paint cost you at the counter?

_____

20 A store owner buys chairs at a cost of $455. They want to make a profit of 12% and must charge the buyer GST. What price should they put on the chairs?

_____

21 A scooter has GST of $42 added to its price.

a What is the value of the scooter before GST? _____

b What is the price after GST has been added? _____

22 An advertisement states **'Save 15%. We pay the GST!'** Investigate this statement.

_____

_____

_____

Standard form

- Standard form is a way of writing either very large or very small numbers without writing all the place holders (0s).
- Standard form is also known as scientific notation.
- A number written in standard form must look like this:

$$4.56 \times 10^n$$

| There must be exactly one non-zero number before the decimal point. | A times sign. | 10 to the power (n) of whatever power is needed to give it the same value as the number in ordinary form. |

Standard form to ordinary numbers

Examples:

1 Numbers bigger than 1

Write 9.26×10^4 as an ordinary number.

Method 1: $9.26 \times 10^4 = 9 . 2\ 6\ 0\ 0 . = 92\ 600$

$4 \Rightarrow$ shift the decimal point 4 places to the **right**.

Some calculators have an 'Exp' button instead.

Method 2: Use your calculator: enter 9.26 4 = 92 600

2 Numbers smaller than 1

Write 9.26×10^{-4} as an ordinary number.

Method 1: $9.26 \times 10^{-4} = 0 . 0\ 0\ 0\ 9 . 2\ 6 = 0.000926$

$-4 \Rightarrow$ shift the decimal point 4 places to the **left**.

Note that for negative powers, the exponent number matches the total number of zeros: $9.26 \times 10^{-4} = 0.000926$

Method 2: Use your calculator: enter 9.26 −4 = 0.000926

Some calculators will write this as '9.26E–04'.

Write these numbers in ordinary form.

1 $1.56 \times 10^4 = $ _____

2 $3.91 \times 10^{-3} = $ _____

3 $5.21 \times 10^5 = $ _____

4 $9.4 \times 10^{-2} = $ _____

5 $2.7 \times 10^6 = $ _____

6 $8.23 \times 10^{-6} = $ _____

Ordinary numbers to standard form

Examples:

1 Numbers bigger than 1

Write 381 000 in standard form.

Step 1: Shift the decimal point to the left until there is exactly one non-zero digit before it. This should give you a number bigger than 1 and smaller than 10.

$$3 . 8 \ 1 \ 0 \ 0 \ 0 . \Rightarrow 3 . 8 1$$

Step 2: Multiply by 10 to the power of the number of places you moved the decimal point.

$$381\ 000 = 3.81 \times 10^5$$

> 381 000 > 3.81 ⇒ **+ve** exponent.

2 Numbers smaller than 1

Write 0.00745 in standard form.

Step 1: Shift the decimal point to the right until there is exactly one non-zero digit before it. Again, this should give you a number bigger than 1 and smaller than 10.

$$0 . 0 \ 0 \ 7 . 4 \ 5 \Rightarrow 7 . 4 5$$

Step 2: Multiply by 10 to the power of the number of places you moved the decimal point.

$$0.00745 = 7.45 \times 10^{-3}$$

> 0.00745 < 7.45 ⇒ **−ve** exponent.

Write these numbers in standard form.

1 41 896 = _____

2 61 = _____

3 956 400 = _____

4 190 = _____

5 8 = _____

6 1 000 000 = _____

7 28.25 = _____

8 400.3 = _____

9 0.5 = _____

10 0.00071 = _____

11 0.0000000006 = _____

12 0.0001011 = _____

13 0.0026 = _____

14 0.000101 = _____

Answer the following questions. Round your answers to 3 sf.

15 A social media app has 3.08×10^6 active daily users. Write this as an ordinary number.

_____

16 The diameter of a hair is about 8.0×10^{-2} mm. Write this as an ordinary number.

_____

17 The wavelength of green light is 5.5×10^{-7} m. Write this as an ordinary number.

_____

18 Giant anteaters consume up to 3.5×10^4 ants and termites in a single day. How many would one eat in a year (365 days)? Write your answer in standard form.

_____

19 There are 1.4 billion insects for each human on the planet. The population of the world is currently 8 billion. Calculate how many insects there are on Earth and write your answer in standard form.

_____

20 The diameter of one hydrogen atom is 1×10^{-12} m. How many hydrogen atoms would fit side by side in one millimetre? Write your answer in ordinary and in standard form.

_____

21 The mass of Earth is 5.97×10^{24} kg. Jupiter is 318 times the mass of Earth. Calculate the mass of Jupiter and write your answer in standard form.

_____

22 Alpha Centauri is 4.4 light years from Earth. One light year is 9.5×10^{12} km. How many kilometres is Alpha Centauri from Earth? Write your answer in standard form.

_____

23 The population of New Zealand is 5.123 million. Each person has, on average, 100 000 hairs on their head. Calculate the total number of head hairs in the population of New Zealand. Write your answer in standard form.

_____

24 The largest virus has a diameter of about 500 nm. One nanometre (nm) is one millionth of a millimetre. Write the width of the largest virus in millimetres. Give your answer in ordinary and in standard form.

_____

Interest

Simple interest

- When you deposit money into the bank, the bank pays you 'interest' for the use of that money.
- When you borrow money, you have to pay 'interest' because you are using the bank's money.
- Simple interest is the same amount paid to you or by you at regular intervals.

Examples:

1 Mika deposits $5400 into a savings account. The bank will pay 4% interest annually. How much interest does she earn after a year?

$$4\% \text{ of } \$5400 = 0.04 \times 5400$$
$$= \$216$$

2 Cameron borrows $1200 off his parents to buy a phone. Cameron will pay them 5% of $1200 in interest each year.

  a How much interest does he pay each year?

$$5\% \text{ of } \$1200 = 0.05 \times 1200$$
$$= \$60$$

  b If it takes him four years to pay the $1200 back, how much interest will he have paid his parents in total?

$$4 \times \$60 = \$240$$

Answer the following questions. All interest rates are annual.

1 Complete the table.

Value of loan	Rate of interest	Interest per year	Number of years	Total debt if none paid back
$5000	6%	$300	4	$5000 + ($300 × 4) = $6200
$800	4%		5	
$21 000	3.5%		6	
$352 000	7.2%		3	

2 Calculate the simple interest earned each year when $18 000 is invested at 5%.

_____

3 Calculate the simple interest earned when $710 000 is invested at 4% for three years.

_____

4 **a** Katrina's grandmother lent her $5500 to buy a car. Katrina paid her 6% in interest each year. How much does she pay each year?

    _____

  b If she pays all the money back after three years, how much will the car have cost her in total?

    _____

5 **a** Materoa invests his $3000 in a savings account with a 4.5% interest rate. How much interest will he earn each year?

    _____

  b How long will he need to leave his money in there to earn $1400 in interest?

    _____

ISBN: 9780170477468

Compound interest

- With compound interest, the interest is added to the amount deposited.
- In the following period, interest is calculated on the total (deposit and previous year's interest).

Amount (value of investment or loan)

rate of interest (decimal), e.g. 5% = 0.05

Formula: $A = P\left(1 + \dfrac{r}{n}\right)^{nt}$

time in years

Principal (how much is invested or borrowed)

number of times interest is compounded per year

Formula if interest applied annually (i.e. once a year, where $n = 1$):

$$A = P(1 + r)^t$$

Examples:

1 Jenny borrows $6500 from her parents and they charge her 5% per year compound interest. This means that she doesn't pay anything back annually, but pays them back the $6500 plus compound interest after three years. How much would she pay back at the end of three years?

End of the first year she owes $6500 + $6500 × 5% = $6825

End of the second year she owes $6825 + $6825 × 5% = $7166.25

End of the third year she owes $7166.25 + $7166.25 × 5% = $7524.56 (2 dp)

Or $6500 × 1.05 × 1.05 × 1.05 = $7524.56 (2 dp)

Or $6500 × 1.05^3 = $7524.56 (2 dp)

2 Ria invested $4000 into a savings account with annual compound interest for six years. At the end of six years, she had $5360.38 in her account. What was the compound interest rate of the account?

$$A = P(1 + r)^t$$

In this situation we need to find r.

$$\$5360.38 = \$4000 \times (1 + r)^6 \quad (\div\ 4000)$$

$$\frac{5360.38}{4000} = (1 + r)^6$$

$$\sqrt[6]{\frac{5360.38}{4000}} = (1 + r)$$

$$1.05 = 1 + r$$

$$r = 0.05$$

So the rate was 5%.

3 Malachi invested $3500 in an account that **compounded monthly** with a rate of 2%. How much would his investment be worth after two years? Describe any assumptions in your answer.

$$A = P\left(1 + \frac{r}{n}\right)^{nt}$$

$$= \$3500\left(1 + \frac{0.02}{12}\right)^{12 \times 2}$$

$$= \$3642.72$$

Assumptions

We assume Malachi didn't withdraw any money over the two years and that the rate remained constant over this time period.

Answer the following questions.

1 Complete the table. All these rates compound annually.

Value of loan	Rate of interest	Number of years	Calculation	Total debt
$5000	3%	4	5000 × 1.03⁴	$5627.54
$200	4%	5		
$18 000	5.5%	2		
$149 500	1.9%	6		

2 Complete the table.

Value of loan	Rate of interest	Number of years	Compound period	Calculation	Total debt
$5000	3%	4	Every six months	$5000\left(1 + \dfrac{0.03}{2}\right)^{2 \times 4}$	$5632.46
$200	4%	5	Monthly		
$18 000	5.5%	2	Fortnightly		
$149 500	1.9%	6	Weekly		

3 $7500 is invested at 6% compound interest, paid annually. Calculate its value at the end of three years.

_____

_____

_____

4 $129 000 is invested at 3.5% compound interest, paid annually. Calculate its value at the end of three years.

_____

_____

_____

5 $16 000 is invested with compound interest paid annually. At the end four years, it was worth $16 576. What was the compound interest rate?

_____

_____

_____

6 At the end of five years, Ria had $7465.38. She had invested her money in an account with compound interest of 4% paid annually for six years. How much was her original investment?

_____

_____

_____

7 Kora has $3000 to invest. She has two offers: scheme one is 4.5% simple interest; scheme two is 3% compound interest paid annually.

a Complete the table to show how much Kora has after one, two, three or four years.

Money withdrawn at the end of	Scheme one	Scheme two
one year		
two years		
three years		
four years		

b What interest rate (2 sf) would Kora require for scheme two to give the same or better result as scheme one at the end of four years?

c Describe any assumptions you have made and how they might affect your answer to **a**.

_____

_____

_____

_____

_____

_____

_____

_____

8 Sione has $15 000 to invest. He has two opportunities: scheme one is 3% compound interest paid annually; scheme two is 3% compound interest, paid every 6 months.

a If he withdraws his investment at the end of three years, how much will each scheme yield?

b What interest rate would Sione require on scheme one to give a better result than scheme two at the end of three years?

Scheme one: _____

_____

_____

Scheme two: _____

_____

_____

c Describe any assumptions you have made and how they might affect your answer.

_____

_____

_____

Algebra

Fundamentals

Simplifying terms

Multiplying and dividing

Unsimplified expressions	Simplified expressions
$a \times a \times a$	a^3
$a \times -5b \times 2a$	$-10a^2b$
$a \div 5$	$\dfrac{a}{5}$ or $\dfrac{1}{5}a$
$\dfrac{32a^2bc}{8ac^2d}$	$\dfrac{4ab}{cd}$

> Because a and c are common to both the top and the bottom, they cancel out.

Adding and subtracting

- Terms can only be added or subtracted if they are 'like' terms.
- 'Like' terms must have exactly the same variables and each variable must be raised to exactly the same power.

Examples:

1 $8\mathbf{a} + a^2 - \mathbf{3a} = 5a + a^2$

2 $7a^2bc - \mathbf{ab^2c} - \mathbf{2ab^2c} = 7a^2bc - 3ab^2c$

Expanding and factorising

Expanding brackets

- In algebra, 'expand' means multiply out all the brackets.
- After expanding, you are expected to collect the like terms in order to simplify the expression.

Examples:

> Be careful!

1 $-3(6x + 4) = -18x - 12$

2 $2x(5 - 7x) = 10x - 14x^2$

3 $8 - 5(6x - 1) = 8 - 30x + 5$
$\qquad\qquad\qquad = -30x + 13$

4 $4(2x - 9) - 3(x + 7) = 8x - 36 - 3x - 21$
$\qquad\qquad\qquad\qquad\quad = 5x - 57$

Factorising

- Factors are terms which are **multiplied** together (rather than added or subtracted), e.g. 2 and 3 are factors of 6 because $2 \times 3 = 6$.
- Expressions with brackets are usually in factorised form.
- When factorising, you must factorise **completely**. There must be no common factor for the terms inside the brackets.

Examples:

Unfactorised (expanded) form	Factorised form
$8x + 12$	$4(2x + 3)$
$7x - 2x^2$	$x(7 - 2x)$
$9x^2 + 3x - 18$	$3(3x^2 + x - 6)$

> There is an unwritten multiply sign here.

Substitution

- When substituting, you replace **variables** with **numbers**.
- It is important to remember **BEDMAS** when doing this.

Examples: Find the value of G when $r = 4$, $h = 3$ and $d = -2$.

1

$$G = \sqrt{7r - h} \div \frac{d}{2}$$

$$= \sqrt{7 \times 4 - 3} \div \frac{-2}{2}$$

$$= \sqrt{25} \div -1$$

$$= 5 \div -1$$

$$= -5$$

2

$$G = \frac{\pi r^2 h}{3}$$

$$= \frac{\pi \times 4^2 \times 3}{3}$$

$$= 16\pi$$

> You may leave π in your answer.

If $b = 5$, $c = 2$, $d = -3$ and $e = -1$, calculate the values of A.

1 $A = 8b + 2d^2$

_____

_____

2 $A = (3c - b)(c - e)$

_____

_____

3 $A = c(d^2 - e)$

_____

_____

4 $A = \sqrt{d^2 + b + c}$

_____

_____

5 $A = \dfrac{b^2 - d^2}{e}$

_____

_____

6 $A = d^3 + b^2$

_____

_____

If $r = 4$ and $h = 5$, calculate the values of V. You may leave π in your answer.

7 $V = h^3 - r^3$

_____

_____

8 $V = \pi r^2 h$

_____

_____

9 $V = \dfrac{4}{3}\pi r^3$

_____

_____

10 $V = r^3 - \dfrac{1}{3}\pi r^3$

_____

_____

Linear equations

Solving linear equations

'Solve' means 'find a value for x'.

Rules:
1. You can do anything you like to an equation as long as you do the **same to both sides**.
2. There should be only **one equals sign** per line.
3. Collect all the variables on one side and numbers on the other side.
4. When you want to get rid of something, perform the **opposite** operation.
5. Your answer should always be in the form $x = \ldots$

Trick: If you need to change the sign of everything, multiply **both** sides by **–1**.

Examples:

1
$$2x - 33 = 19 \qquad (+\ 33)$$
$$2x = 19 + 33 \qquad (\div\ 2)$$
$$x = \frac{19 + 33}{2}$$
$$x = 26$$

2
$$9 + 7x = 3x - 7 \qquad (-\ 9,\ -\ 3x)$$
$$7x - 3x = -9 - 7 \qquad (\text{simplify})$$
$$4x = -16 \qquad (\div\ 4)$$
$$x = -4$$

3
$$2(x + 3) = 12 - x \qquad (\text{expand})$$
$$2x + 6 = 12 - x \qquad (-\ 6,\ +\ x)$$
$$3x = 6 \qquad (\div\ 3)$$
$$x = 2$$

4
$$\frac{5x + 2}{2} = \frac{3x - 8}{3} \qquad (\times\ 6)$$
$$3(5x + 2) = 2(3x - 8) \qquad (\text{expand})$$
$$15x + 6 = 6x - 16 \qquad (-\ 6,\ -\ 6x)$$
$$9x = -22 \qquad (\div\ 9)$$
$$x = \frac{-22}{9}$$

> Leave answers in fraction form rather than rounding a decimal.

Solve the following.

1
$$98 - 3x = 4x - 7$$

2
$$2(x - 9) = 27 - 5x$$

3
$$\frac{6x - 6}{2} = 8 - x$$

4
$$4(x - 3) = -2(7x - 3)$$

ISBN: 9780170477468

5 $$\frac{2x - 6}{4} = \frac{5x + 6}{3}$$

6 $$15 = \frac{9x}{2} - 3$$

7 $2(x - 16) + 1 = 41$

8 $6x - 1 - x = 24 - 3x$

9 $$\frac{1}{2}(8x - 12) = 19$$

10 $$\frac{2x - 1}{3} = \frac{7}{12}$$

11 $$\frac{2(x + 8)}{3} = 17$$

12 $6(11 - 5.75x) = -3(21x + 16)$

13 $7(2x - 3) - (4 - x) = 5(6 - x)$

14 $$\frac{6x - 3}{x} = 15$$

15 $$\frac{3}{5}(x - 8) = \frac{6x}{5}$$

16 $$\frac{2x - 3}{3} + \frac{3x + 4}{5} = 15$$

Forming and solving linear equations

- **Define** your **variable** using the first letter of the word whose value you need to find.
- If you need to find several values, it is usually easiest to select the smaller value as your variable.
- Often it is useful to draw a diagram.

Some words used for the four basic operations:

+	plus, total, more, and, add(ed), sum, increased by, greater than	–	subtract(ed), less, decreased by
x	of, times, multiplied by, product	÷	divided by, shared between, per, out of

- 'Double' and 'twice' both mean multiply by two.

Examples:

1 Seven people go skiing. Adult tickets cost $159 and children's tickets cost $99. The total cost was $873. Write an equation and use it to help you find the numbers of adults and children who went skiing.

Let *a* represent the **number of adults**. — Always define your **variable**.

Then: $159a + 99(7 - a) = 873$ — number of children = 7 – number of adults

$$159a + 693 - 99a = 873$$
$$60a + 693 = 873$$
$$60a = 180$$
$$a = 3$$

∴ Three adults and four children went skiing. — Always answer the question in a **sentence**.

2 A woman was 27 when her son was five years old. When will she be triple her son's age?

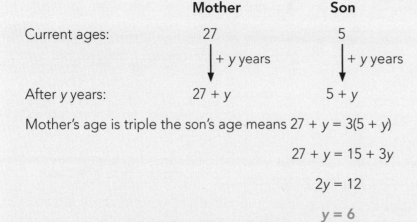

	Mother	Son
Current ages:	27	5
	+ y years	+ y years
After y years:	27 + y	5 + y

Mother's age is triple the son's age means $27 + y = 3(5 + y)$

$$27 + y = 15 + 3y$$
$$2y = 12$$
$$y = 6$$

So, in 6 years' time she will be triple his age.

Check: She will be 33 and he will be 11. ✓

Write and solve equations to find the unknown values.

1 The long sides of an isosceles triangle are 7 cm longer than the shortest side. The perimeter of the triangle is 29 cm. What are the dimensions of the triangle?

_____

_____

_____

_____

_____

_____

2 Kelsey bought five large containers and three small ones for a total of $22. The large containers cost $2 more than the small ones. Write an equation and use it to help you find the price for each size of container.

_____

_____

_____

_____

_____

_____

3 A rectangle has short sides that are 3 cm shorter than the long sides. It has a perimeter of 42 cm. Write an equation relating the perimeter to the lengths of the sides, and use it to calculate the dimensions of the rectangle.

_____

_____

_____

_____

_____

4 Twelve people go to the circus. Adult tickets cost $45 and children's tickets cost $18. The total cost was $297. Write an equation and use it to help you find the numbers of adults and children who went to the circus.

_____

_____

_____

_____

_____

5 Kurt is seven years old; he was born when his dad was 28. When will his dad be double Kurt's age?

_____

_____

_____

_____

Plotting linear equations

- There are several ways to plot linear equations.
- Tables will always work for any equation.
- Always plot **at least three points** and connect them with a **ruled line**.
- Always rule the line to the edges of the axes.

Example: Plot the line given by the equation $y + 2.5x = 12$.

Rearrange: $y = -2.5x + 12$

> It is easiest to complete the table if the equation is in the form
> $$y = \ldots$$

Fill in the table, plot the points, then connect them with a straight line.

x	Calculation	y	Coordinates
0	$-2.5 \times 0 + 12$	12	(0, 12)
1	$-2.5 \times 1 + 12$	9.5	(1, 9.5)
2	$-2.5 \times 2 + 12$	7	(2, 7)
3	$-2.5 \times 3 + 12$	4.5	(3, 4.5)
4	$-2.5 \times 4 + 12$	2	(4, 2)
5	$-2.5 \times 5 + 12$	-0.5	(5, -0.5)

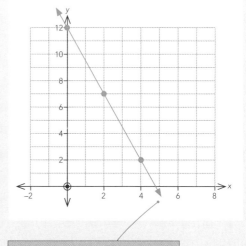

> Sometimes you will get points that don't fit on the axes you are given. If this happens, just leave the point out.

> Notice that line continues beyond the coordinates.

Complete the tables and draw the graphs for the following equations.

1 $y = -2x + 15$

x	y
0	
1	
2	
3	
4	
5	
6	

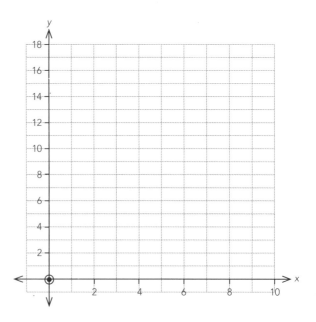

2 $y = -\dfrac{1}{2}x + 7$

x	y
0	
2	
4	
6	
8	
10	

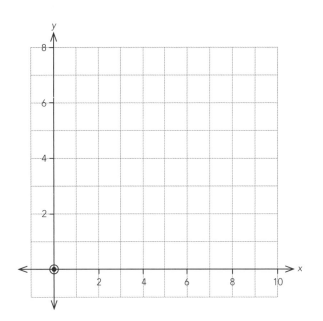

3 $2x - y = -8$

x	y
0	
1	
2	
3	
4	
5	

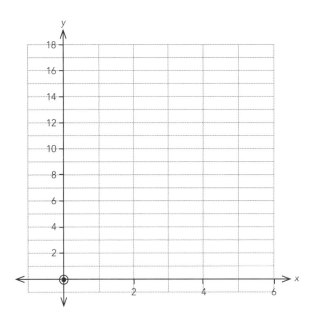

4 $12x = 32 - 2y$

x	y
0	
1	
2	
3	

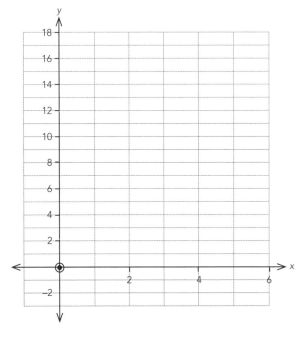

ISBN: 9780170477468

The gradient of a line

- The gradient is the steepness or slope of a line.

These lines all have positive gradients.

These lines all have negative gradients.

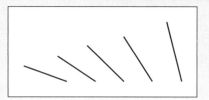

The steeper the line, the bigger the gradient.

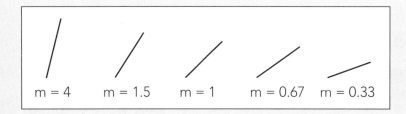

m = 4   m = 1.5   m = 1   m = 0.67   m = 0.33

The gradient is calculated using the formula $m = \dfrac{\text{change in } y}{\text{change in } x}$ or $\dfrac{\text{rise}}{\text{run}}$.

The easiest way to do this is to draw a right-angled triangle on the line.

$$m = \dfrac{\text{rise}}{\text{run}}$$
$$= \dfrac{5}{6}$$

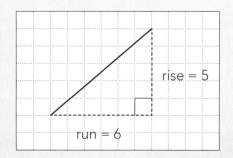

rise = 5

run = 6

This time the gradient is *negative*.

$$m = -\dfrac{\text{rise}}{\text{run}}$$
$$= -\dfrac{3}{7}$$

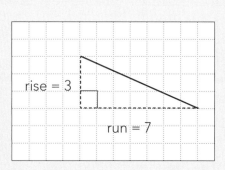

rise = 3

run = 7

Horizontal lines

$$m = \frac{rise}{run}$$

$$= \frac{0}{10}$$

$$= 0$$

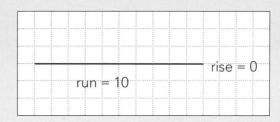

run = 10   rise = 0

Vertical lines

$$m = \frac{rise}{run}$$

$$= \frac{8}{0}$$

$$= undefined$$

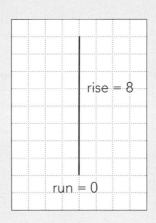

rise = 8

run = 0

1 Write the gradients of these lines.

18

a   m = _____

b   m = _____

c   m = _____

d   m = _____

e   m = _____

f   m = _____

g   m = _____

h   m = _____

i   m = _____

j   m = _____

Using features of straight lines to write equations

- Graphs of lines are a way of illustrating relationships.
- The gradient and the y intercept enable us to write equations and interpret contexts.

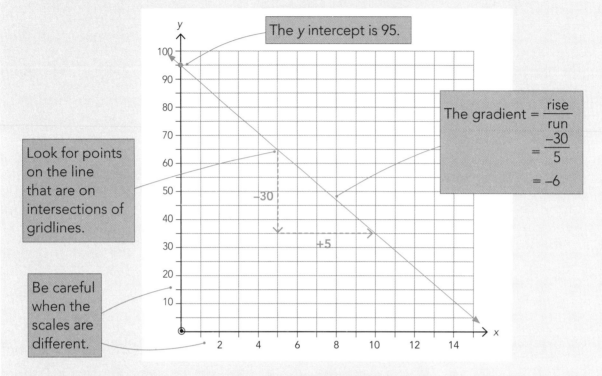

The y intercept is 95.

Look for points on the line that are on intersections of gridlines.

$$\text{The gradient} = \frac{\text{rise}}{\text{run}}$$
$$= \frac{-30}{5}$$
$$= -6$$

−30

+5

Be careful when the scales are different.

To write an equation for this linear function, use the structure **y = mx + c**.

m is the **gradient**.   **c** is the **y intercept**.

So for this example, the equation is **y = −6x + 95**.

Identify the features and write the equation for these graphs.

1

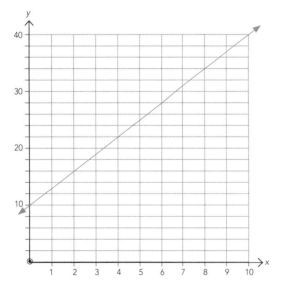

y intercept = _____ Gradient = _____

Equation y = _____x ____ _____
                         ↑
                        sign

2

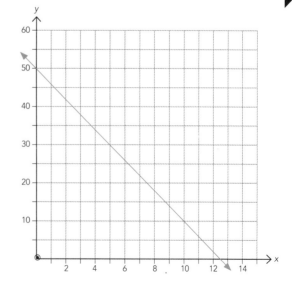

y intercept = _____ Gradient = _____

Equation y = _____x ____ _____
                         ↑
                        sign

3

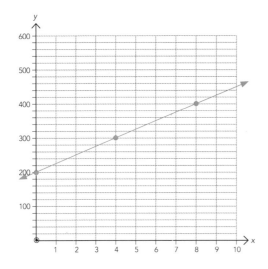

y intercept = _____ Gradient = _____

Equation y = _____

4

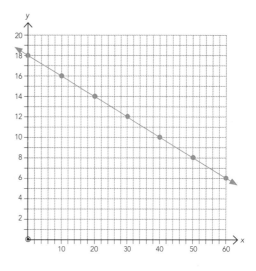

y intercept = _____ Gradient = _____

Equation y = _____

5

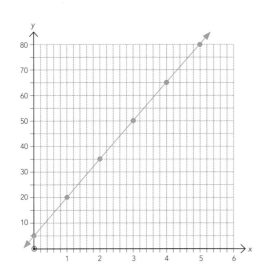

y intercept = _____ Gradient = _____

Equation y = _____

6

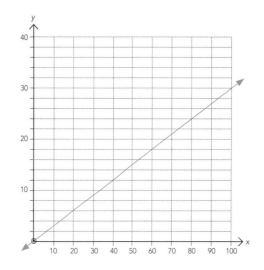

y intercept = _____ Gradient = _____

Equation y = _____

7

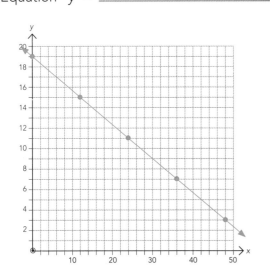

y intercept = _____ Gradient = _____

Equation y = _____

8

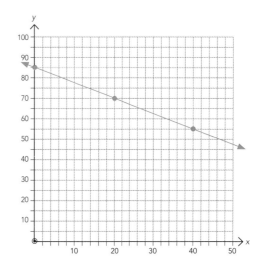

y intercept = _____ Gradient = _____

Equation y = _____

Equations of vertical, horizontal and parallel lines

Equations of vertical lines

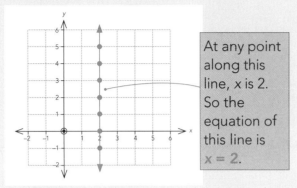

At any point along this line, x is 2. So the equation of this line is $x = 2$.

Equations of horizontal lines

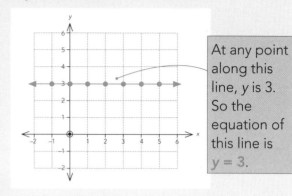

At any point along this line, y is 3. So the equation of this line is $y = 3$.

Be **very careful** when writing these because the line $x = \ldots$ is parallel with the **y**-axis, and the line $y = \ldots$ is parallel with the **x**-axis.

Parallel lines have the same gradient.

Example:

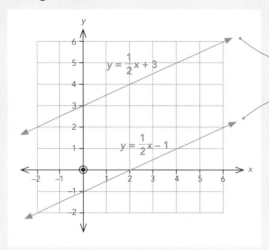

$$y = \frac{1}{2}x + 3$$

$$y = \frac{1}{2}x - 1$$

$m = \dfrac{1}{2}$ for both equations.

Write the equations for these lines.

1

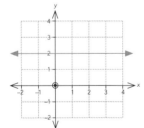

_____

2

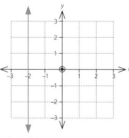

_____

The lines on each graph are parallel. Write equations for the dashed lines.

3

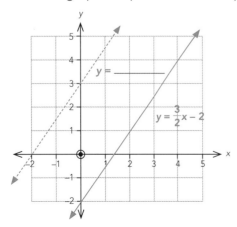

$y =$ _____

$$y = \frac{3}{2}x - 2$$

4

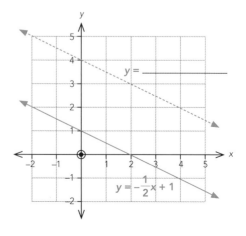

$y =$ _____

$$y = -\frac{1}{2}x + 1$$

 ISBN: 9780170477468

Writing equations given two points

- Sometimes you don't have a graph or finding the y intercept is difficult, so you can't easily use $y = mx + c$.
- In these cases, you will need to calculate the gradient (m) of the line using the formula:

$$m = \frac{rise}{run} = \frac{y_2 - y_1}{x_2 - x_1}$$

- Then use the gradient (m) along with the coordinates of one point (x_1, y_1) in the formula:

$$y - y_1 = m(x - x_1)$$

Example: Calculate the equation for the line that connects (2, 39) and (12, 14).

Step 1: Name the coordinates: (2, 39)   and   (12, 14).

x_1 y_1 x_2 y_2

Step 2: Calculate the gradient:

$$m = \frac{y_2 - y_1}{x_2 - x_1} = \frac{14 - 39}{12 - 2}$$

$$= \frac{-25}{10}$$

$$= -2.5$$

Step 3: Calculate the equation:

$$y - y_1 = m(x - x_1)$$
$$y - 39 = -2.5(x - 2) \qquad \text{(expand and add 39)}$$
$$y = -2.5x + 5 + 39$$
$$y = -2.5x + 44$$

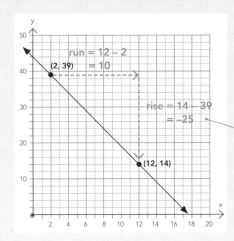

The calculation of the gradient is the same as you would do on a graph.

Calculate the equations for the lines that connect these points.

1   (4, 19) and (10, 43)

_____

_____

_____

_____

2   (4, 15) and (17, 28)

_____

_____

_____

_____

3 (2, 17) and (9, 3)

_____

_____

_____

_____

4 (12, 1) and (18, 4)

_____

_____

_____

_____

5 (4, 35) and (8, 38)

_____

_____

_____

_____

6 (6, 40) and (14, 36)

_____

_____

_____

_____

7 (4, 14) and (12, 42)

_____

_____

_____

_____

8 (0, 73) and (14, 38)

_____

_____

_____

_____

9 An electrician's labour charges were $142.50 for a job that took an hour and a half, and $240 for a job that took three hours.

    a Calculate the equation that represents his charging structure.

    _____

    _____

    _____

    b What did he charge for his travel time? $_____

    c What was his hourly rate? $_____

Applications

- The value of **c (the y intercept)** tells you where the graph **cuts the y-axis**. This is often the 'starting point' for a relationship because it tells you the **value of y when x is zero**.
- The value of **m (the gradient)** tells you the **rate of change** of the variable on the y-axis compared with that on the x-axis.

Example: Jess is an electrician. The amounts she charges for a call-out are shown on the graph.

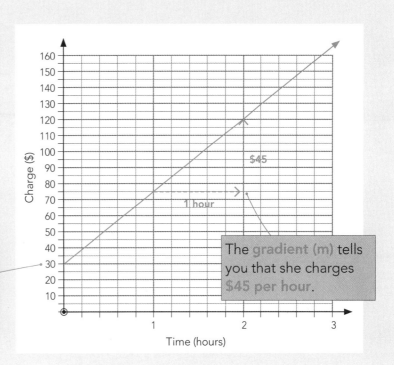

The **y intercept (c)** tells you that her **minimum charge** (to cover travel time) is **$30**.

The **gradient (m)** tells you that she charges **$45 per hour**.

Answer the following questions.

1 Nisha is saving to go to a sports tournament that will cost her $560. The graph shows the total amount (S) that Nisha saves.

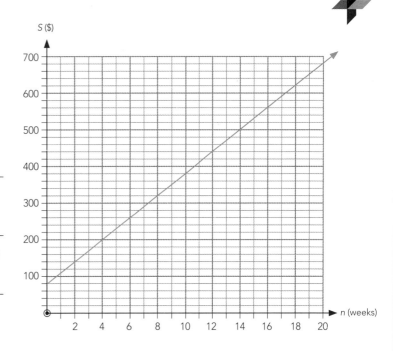

a How much money did Nisha have to start with?

_____

b How is this shown on the graph?

_____

c How much does Nisha save each week?

_____

d How is this shown on the graph?

_____

e Write an equation showing the relationship between the total amount Nisha has saved (S) and time (n):

S = _____

f After how many weeks will Nisha have the $560 that she needs? Show how you found this on the graph.

_____ weeks

2   Hamish hires a Lemon electric scooter. The graph shows how much it costs him.

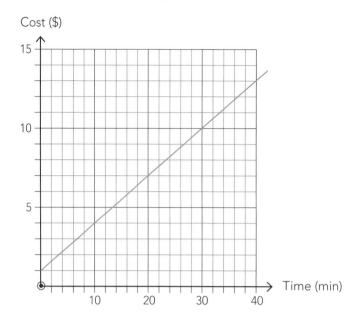

a   How much money did it cost to unlock the scooter? $_____

b   How is this shown on the graph? _____

c   How much do Lemon scooters cost per minute? $_____

d   How is this shown on the graph? _____

e   Write an equation showing the relationship between the total cost of the scooter ride
(C) and time (t).

_____

f   Hamish thinks he will use the scooter for 20 minutes. How much will this cost him?

$_____

g   He is charged $8.20. How long did he have the scooter for?   _____ min

h   Banana scooters cost $2 to unlock and cost the same amount per minute as Lemon scooters.
How would the graph for Banana scooters differ from the graph for Lemon scooters?

_____

i   What would be the same for the two graphs?

_____

j   Write an equation for Banana scooters showing the relationship between the total cost
of the scooter ride (C) and time (t).

_____

k   Orange scooters cost $6 for 10 minutes of use and $12 for 40 minutes of use. Write an
equation to represent this relationship.

_____

3 Hemi is selling bags of lollies at his local fair. The graph shows the number of bags sold and his profit.

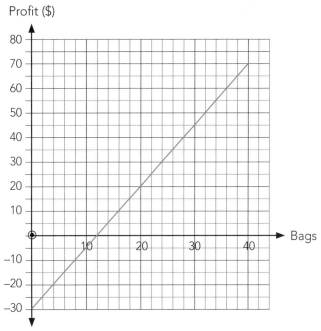

Profit ($)

Bags

a How much does he sell each bag for? $_____

b How is this shown on the graph? _____

c How much does it cost him to buy each bag of lollies? $_____

d How much profit will he make if he sells 30 bags? $_____

e Write an equation showing the relationship between profit (*P*) and the number of bags sold (*b*).

_____

f How many bags will he need to sell to break even (make $0 dollars)?

$_____

g Explain his situation if he sold only 8 bags of lollies.

_____

_____

h His sister told him that he shouldn't have been so greedy, and he should have sold the bags for $2 each. Add this relationship to the graph and explain in detail how this would have changed the relationship between his profit and the number of bags sold. Include an equation in your answer.

_____

_____

_____

4 A spa pool contains 150 L of water. It is then filled up at a constant rate of 25 L per minute.

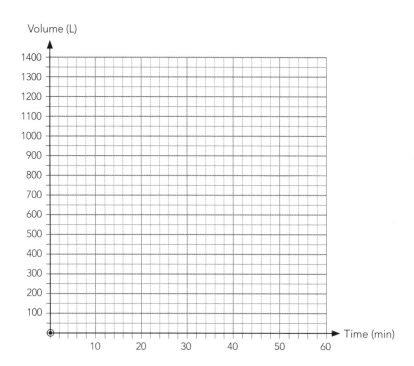

a On the axes, draw a line showing the filling of the spa.

b How many litres are in the spa after 38 minutes? _____ L

c Mark this point on the graph.

d After how long will the spa contain 1350 litres? _____ min

e Write an equation showing the relationship between the amount of water (*V*) and time (*t*).

_____

f The spa can hold 1550 litres of water. How long will it take to fill? _____ min

g Write the equation that would represent this relationship if there had been 300 L of water in the spa at the start.

_____

h Describe any differences or similarities between a line representing this relationship and the one that you drew in **a**.

_____

_____

_____

i On another day, the spa contained 350 L after 10 minutes of filling and 1100 L after 60 minutes. Write an equation to represent this relationship.

_____

5 Aiko is walking to a lake to meet a friend. The graph shows how far she is from the lake and the time since she started walking.

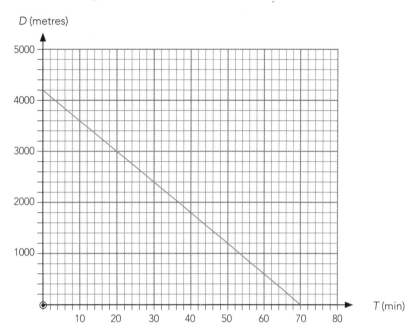

a How far does she need to walk in order to reach the lake? _____ km

b How is this shown on the graph? _____

c How fast is Aiko walking? _____ m/min = _____ km/hour

d How is this shown on the graph? _____

e How far has she walked after 15 minutes? _____ m

f How long does it take her to walk 3 km? _____ min

g Write an equation for Aiko's walk showing the relationship between her distance from the lake and time.

_____

h After half an hour she realises that she will be late and she increases her pace to a steady 4.8 km/hour. Add the line showing this to the graph, and write its gradient.

_____

_____

_____

i What difference will this make to her arrival time at the lake?

_____

_____

Piecewise functions

- A **piecewise** function is a graph with several **different** rules (formulas).
- The rule for each section only applies to a certain range of x values.
- Piecewise functions occur when there are **different** rules for finding y, depending on the value of x.

Examples:

1 There are three functions on these axes.

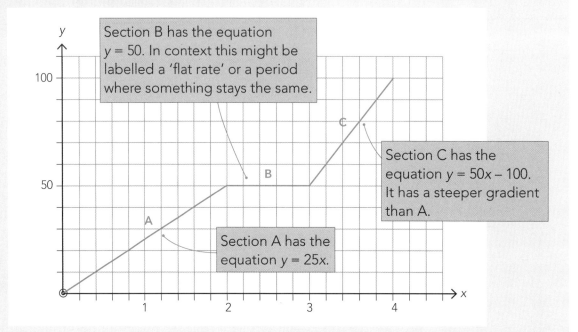

Section B has the equation y = 50. In context this might be labelled a 'flat rate' or a period where something stays the same.

Section C has the equation $y = 50x - 100$. It has a steeper gradient than A.

Section A has the equation $y = 25x$.

2 Yelana is selling hams. The price of a ham depends on its mass. A ham which is less than 1 kg costs $15. A ham which is 1 kg or more, but less than 2 kg, costs $20. The price increases in $5 steps for each kilogram.

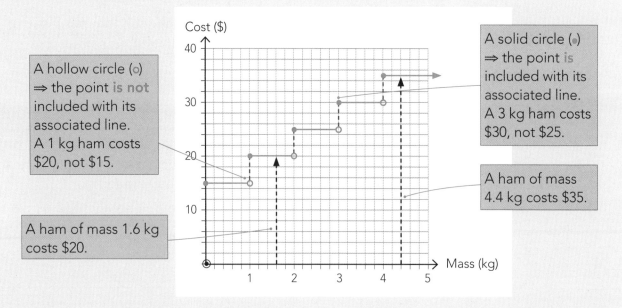

A hollow circle (o) ⇒ the point **is not** included with its associated line. A 1 kg ham costs $20, not $15.

A solid circle (•) ⇒ the point **is** included with its associated line. A 3 kg ham costs $30, not $25.

A ham of mass 4.4 kg costs $35.

A ham of mass 1.6 kg costs $20.

You can use the graph to work out the cost for any mass of ham.

Answer the following questions.

1   **a**   A cellphone company uses the functions below to determine the cost for phone calls.

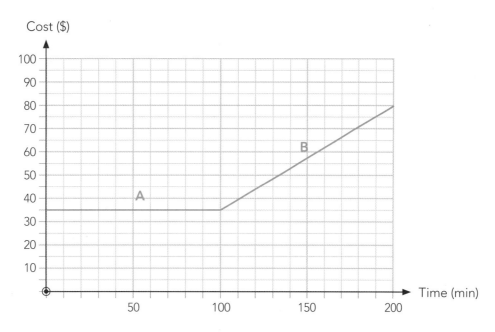

a   Section A shows a set charge of $ _____ for _____ minutes or fewer.

b   If a customer goes over 100 minutes of calls, they are charged _____ per minute.

2   A new bike hire company charges users as shown in the graph below.

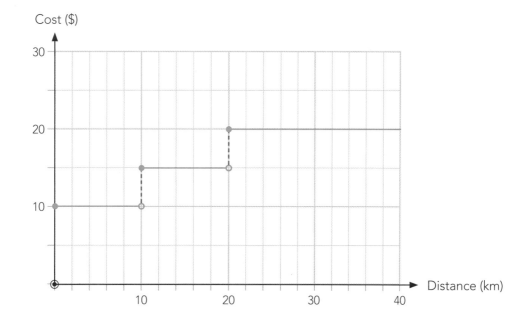

a   How much will it cost to hire a bike and travel 9 km? _____

b   How much will it cost to hire a bike and travel 20 km? _____

3 The graph shows Tama's bike trip from home to his netball game and back.

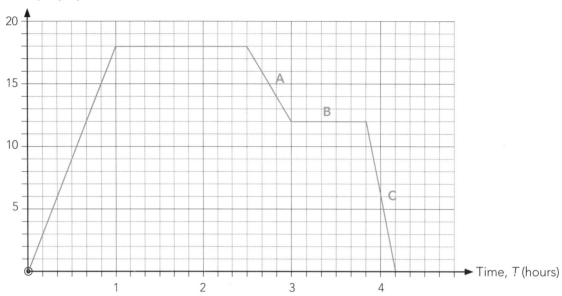

Distance, *D* (km)

Time, *T* (hours)

a Write equations for the following sectors of his journey.
During the first hour: _____

Between one hour and two and a half hours after leaving home: _____

b It took him one hour to bike to his netball game.

How far was the game from his home? _____ km

c How fast was he cycling? _____ km/h

d How long did he spend at his netball game? _____ h _____ min

e How long did it take him to get home from the netball courts? _____ h _____ min

f The trip home had three stages. Write the gradient for each. A: _____

B: _____

C: _____

g Describe in detail what the graph tells you about the trip home.

_____

_____

_____

_____

_____

 ISBN: 9780170477468

4 Oscar the musician charges customers as per the box below.

Time, T (min)	Cost, C ($)
0 – 119	$160
120 – 239	$230
240 – 300	$330

a Show these charges on the graph.

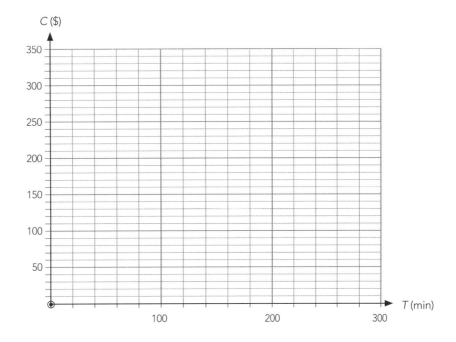

b Find the equation for each part of the graph. $0 \leq T$ (min) < 120: _____

$120 \leq T$ (min) < 240: _____

$240 \leq T$ (min) ≤ 300: _____

c How much does he charge for a job that takes an hour? _____

d How much does he charge for a job that takes four hours? _____

e Describe any assumptions you have made and how they might affect your answer.

_____

_____

_____

Simultaneous equations

- 'Simultaneous' means '**at the same time**'.
- For simultaneous equations, you need to be able to solve two equations that are both true 'at the same time'.
- Because there are two equations, there are also two variables.
- There are two methods for solving these, and which one you use depends on how the equations are structured.

Solving simultaneous equations

1 Substitution
- Substitution is easiest where one of the equations is expressed as $x = ...$ or $y = ...$.
- As the name suggests, you substitute the $x = ...$ or $y = ...$ into the other equation.
- You should number each equation, and say what you are doing at each step.

Example: Solve the equations $y = 3 - x$ and $3y + x = 5$.

Number the equations.

$$y = 3 - x \quad ①$$
$$3y + x = 5 \quad ②$$

Substitute ① into ②: $\quad 3(3 - x) + x = 5$

From equation ① we know that **y** and **3 − x** are equal.

$$9 - 3x + x = 5$$
$$9 - 2x = 5$$
$$-2x = -4$$
$$x = 2$$

Substitute for x in ①: $\quad y = 3 - 2$
$$y = 1$$

You can also write the solution as (2, 1).

2 Elimination
- Elimination is easiest when the two equations have the same structure.

For example:

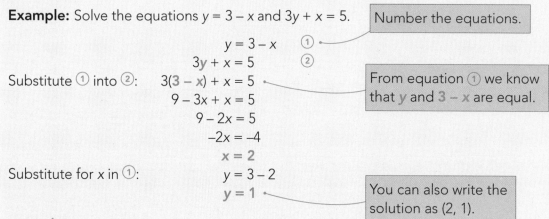

- You solve these by multiplying one or both equations by a constant in order to make either the x terms or the y terms the **same size** but with **different signs**.
- Then **add** the two equations to produce an equation with one variable only.
- You should number each equation, and say what you are doing at each step.

Example: Solve the equations $x + 3y = -13$ and $4x - 7y = 43$.

$$x + 3y = -13 \quad ①$$
$$4x - 7y = 43 \quad ②$$

Multiply ① by −4: $\quad -4x - 12y = 52 \quad ③$

It is a good idea to check your answer by substituting into the other equation. ② $4(2) - 7(-5) = 43$ ✓

Add ② and ③: $\quad -19y = 95$
$$\therefore \quad y = -5$$

Substitute for y in ①: $x + (3) \times (-5) = -13$
$$\therefore \quad x = 2$$

Solve the following simultaneous equations.

1
$$x = 4 + y$$
$$2x + 3y = 23$$

_____

_____

_____

_____

2
$$2x + y = 22$$
$$5x + y = 10$$

_____

_____

_____

_____

3
$$2y + 3x = 20$$
$$y = x + 5$$

_____

_____

_____

_____

4
$$x - y = 5$$
$$x - 2y = 8$$

_____

_____

_____

_____

5
$$4x + 2y = 16$$
$$2x + 3y = 12$$

_____

_____

_____

_____

_____

6
$$3x + 2y = 14$$
$$y + 2x = 9$$

_____

_____

_____

_____

_____

7
$$x + 2y = 12$$
$$8 - 3x = y$$

_____

_____

_____

_____

_____

8
$$4x - 3y = 10$$
$$6x + 4y = 49$$

_____

_____

_____

_____

_____

Solving simultaneous equations using graphs

Combine the skills that you have learnt so far in order to answer these questions.

1 Cru's Taxis' charges are graphed on the axes.

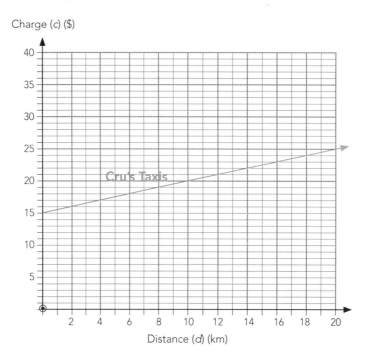

a Cru's Taxis charges a fixed price plus an amount for every kilometre travelled.

How much is the fixed price? $_____

b How much does Cru's Taxis charge per kilometre? $_____

c Cat's Cabs has a similar business model. Cat's Cabs charges $5 plus $1.50 per kilometre travelled. Plot this relationship on the axis.

d Write equations for the amount charged by both companies:

Cru's Taxis: _____

Cat's Cabs: _____

e How much does each company charge for an 8 km ride?

Cru's Taxis: $_____

Cat's Cabs: $_____

f If a customer has only $23, how far would they be able to travel with each company?

Cru's Taxis: _____ km

Cat's Cabs: _____ km

g For what distance do both companies charge the same amount? _____ km

How much do they charge for that trip? $_____

2   Sarah and her friend Aroha love to go to watch movies.

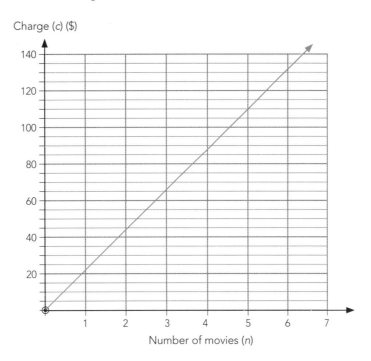

Charge (*c*) ($)

Number of movies (*n*)

a   Sarah joins the loyalty programme, which means each movie costs her only $17. However, she has to pay a $20 annual fee. Plot this relationship on the axis.

b   Aroha does not belong to the loyalty programme. The relationship between the number of movies she goes to and the cost is plotted on the graph.

How much does a ticket to one movie cost? $_____

c   Write equations for the movie costs for each friend: Sarah: _____

Aroha: _____

d   If they go to six movies, who pays the most and by how much?

_____

_____

e   How many movies will Sarah need to see to make her loyalty card worthwhile? _____
Explain how you reached this answer.

_____

_____

_____

f   The movie theatre would like more loyalty card customers. Describe two ways that they could adjust their current model to attract more customers.

_____

_____

3 Jai wants to book a clown for his daughter's birthday party.

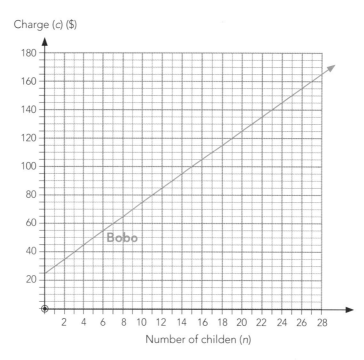

Charge (*c*) ($)

Number of children (*n*)

a Jojo the clown charges a $60 booking fee and then $2 per child. Plot this on the axes.

b Bobo the clown's prices are plotted on the axes.

How much is Bobo's booking fee? $_____

c How much does Bobo charge per child? $_____

d Write equations for each charging structure.   Bobo: _____

Jojo: _____

e Sometimes the coordinates for the point of intersection cannot be read accurately from your graph. Solve the equations simultaneously to find these coordinates and explain what they mean.

_____

_____

_____

_____

f Jojo is concerned that he is losing business to Bobo, especially for smaller parties. Suggest a change to his current fee structure that would make Jojo more competitive.

_____

_____

_____

_____

4   Gia and Flo are in a 50 km cycle race. Gia knows she can cycle at a speed of 26 km per hour. Flo is a little faster at 30 km per hour. Unfortunately, Flo gets a flat tyre on the start line and starts the race 12 minutes after Gia.

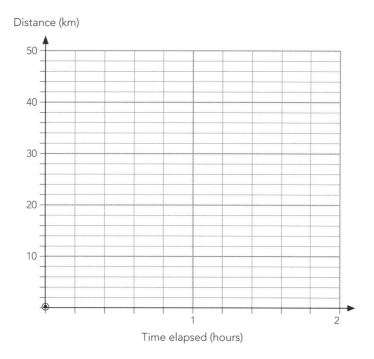

a   Plot the progress of both racers on the axes.

b   Write equations for each of the two racers.   Gia: _____

Flo: _____

c   The coordinates for the point of intersection will not be easily read from your graph. Solve the equations simultaneously to find these coordinates.

_____

_____

_____

d   Check that these coordinates fit with your graph, and write a description of Gia's and Flo's progress during the race, including their finishing times.

_____

_____

_____

e   Describe any assumptions made in answering this question.

_____

_____

_____

Forming and solving simultaneous equations

- Read what the question asks for. Define your variables using their first letters.
- Write two equations for each question, then solve them.

Example: Yelena has twice as many lollies as Xavier. If Xavier had another 10 lollies, he would have three times as many as Yelena. How many lollies does each person have?

Call the number of lollies Yelena has **y**.

Always define your **variables**.

Call the number of lollies Xavier has **x**.

Yelena (**y**) has (=) twice as many lollies as Xavier (**2x**).

$$y = 2x \quad ①$$
$$x + 10 = 3y \quad ②$$

If Xavier had another 10 lollies (**x + 10**), he would have (=) three times as many as Yelena (**3y**).

Substitute ① into ②:
$$x + 10 = 3(2x)$$
$$x + 10 = 6x \quad (-x)$$
$$10 = 5x$$
$$x = 2$$

Substitute for x in ①:
$$y = 2(2)$$
$$y = 4$$

∴ Xavier has 2 lollies and Yelena has 4 lollies.

Always answer the question in a sentence.

Write and solve simultaneous equations to find the unknown values.

1 Ed went shopping and bought three capsicums and two avocados. They cost him $14.17. Sumi bought one capsicum and five avocados. They cost her $15.99. Calculate the price of capsicums and avocados.

2 Taylor and Sam have $85 in total. Taylor has seven dollars more than twice what Sam has. How much money does each of them have?

3 Victor and Lee have a combined age of 58. Two years ago, Victor was three times the age Lee is now. How old are Victor and Lee?

_____

_____

_____

_____

_____

_____

_____

_____

_____

4 Tickets to the fair cost $3 for children and $15 for adults. The fair owners sell 144 tickets and take in $1524. How many children and adults attended the fair?

_____

_____

_____

_____

_____

_____

_____

_____

_____

5 Miley has three times as many Excellence credits as Achieved credits. The number of Excellence credits she has is double her number of Achieved credits plus three. Calculate how many Achieved and Excellence results Miley has.

_____

_____

_____

_____

_____

_____

_____

_____

6 In a survey, there were 12 times as many right-handed students as left handed. One ninth of the number of right-handed students is two more than the number of left-handed students. How many right-handed and left-handed students were surveyed?

_____

_____

_____

_____

_____

_____

_____

_____

Linear inequalities

- Inequations are equations that have **<**, **≤**, **>** or **≥** signs.
  - **<** means 'is less than'
  - **>** means 'is greater than'
  - **≤** means 'is less than or equal to'
  - **≥** means 'is greater than or equal to'
- You solve an inequation in exactly the same way as you solve an equation **except**:
  - if you need to **multiply or divide** the equation by a **negative** number
  - **or** if you **swap the sides**,
  - **then** **you must reverse the sign**.

Examples:

1 $9x \leq 72$ $(\div 9)$

$$x \leq \frac{72}{9}$$

$$x \leq 8$$

This means that x is less than or equal to 8.

2 $6x + 3 > -39$ $(- 3)$

$$6x > -42 \qquad (\div 6)$$

$$x > -7$$

This means that x is greater than -7.

3 $32 < 2 + 5x$ $(- 2)$

$$30 < 5x$$

$$5x > 30 \qquad (\div 5)$$

$$x > 6$$

> Swap the sides ⇒ reverse the sign.

This means that x is greater than 6.

> Divide by a **negative** number ⇒ **reverse the sign**.

4 $\dfrac{3 - 2x}{5} > \dfrac{x + 3}{2}$ $(\times 10)$

$$6 - 4x > 5x + 15 \qquad (- 6)$$

$$-4x > 5x + 9 \qquad (- 5x)$$

$$-9x > 9 \qquad (\div -9)$$

$$x < -1$$

This means that x is less than -1.

Solve the following and complete the sentence.

1 $7x > 63$

_____

_____

This means that x is _____ than ____

2 $4x + 1 \leq 49$

_____

_____

This means that x is _____ than or

equal to ____

3 $3 + 7x < 31$

_____

_____

This means that x is _____ than ____

4 $9x - 1 \geq x + 23$

_____

_____

This means that x is _____

5

$$-3x < 39$$

_____

_____

This means that x is _____

6

$$2(x - 1) \geq x$$

_____

_____

This means that x is _____

7

$$5 - 2x > 43$$

_____

_____

This means that x is _____

8

$$6x + 4 \geq 9x - 1$$

_____

_____

This means that x is _____

9

$$\frac{3x}{4} > x + 2$$

_____

_____

This means that x is _____

10

$$\frac{10 - 3x}{2} > \frac{x + 4}{3}$$

_____

_____

This means that x is _____

11 The width of a rectangular cake is a third of its length. Trudi has a 1.52 m long ribbon that will be tied around the perimeter of the cake. The bow will need 40 cm. Calculate the maximum dimensions of the cake.

_____

_____

_____

_____

12 Terry has budgeted to spend a maximum of $200 on plants for his garden. He will plant kōwhai trees, which cost $18, and hebes, which cost $9. He wants to have at least three hebes for every kōwhai tree, and as many kōwhai as possible. How many hebes and kōwhai trees can he afford?

_____

_____

_____

_____

Rearrangement of expressions

- It is often useful to rearrange an expression so a different variable becomes the subject.
- Use normal equation-solving rules to isolate the subject.

Rules:
1. You can do anything you like to an equation as long as you do the **same to both sides**.
2. There should be just **one equals sign** per line.
3. Collect the variable required on one side and put everything else on the other side.
4. When you want to get rid of something, perform the **opposite** operation.
5. Your answer should always be in the form $v = ...$

Trick: If you need to change the sign of everything, multiply **both** sides by –1.

Examples:

1. Make T the subject of $I = \dfrac{PRT}{100}$. (Move the variable to the left.)

$$\dfrac{PRT}{100} = I$$ (x 100)

$$PRT = 100\,I$$ (÷ PR)

$$T = \dfrac{100\,I}{PR}$$

2. Make R the subject of $A = \pi(R^2 - r^2)$. (Move the variable to the left.)

$$\pi(R^2 - r^2) = A$$ (Expand the brackets.)

$$\pi R^2 - \pi r^2 = A$$ (Move the term without the variable to the right.)

$$\pi R^2 = A + \pi r^2$$ (÷ π)

$$R^2 = \dfrac{A}{\pi} + r^2$$ (Take the square root of both sides.)

$$R = \sqrt{\dfrac{A}{\pi} + r^2}$$

Answer the following questions.

1. The formula for the perimeter of a circle is $P = 2\pi r$. Make r the subject of this formula.

_____

_____

_____

2. The formula for velocity is $v = \dfrac{d}{t}$. Make d the subject of the formula.

_____

_____

_____

3 The formula for the area of a circle is $A = \pi r^2$. Make r the subject of this formula.

_____

_____

_____

_____

4 Rearrange $y = mx + c$ so x is the subject of the formula.

_____

_____

_____

_____

5 The formula for the sum (S) of the internal angles of a polygon is $S = 180(n - 2)$. Make n the subject of the formula.

_____

_____

_____

_____

6 The formula for the volume of a cone is $V = \dfrac{1}{3}\pi r^2 h$. Make r the subject of this formula.

_____

_____

_____

_____

7 The formula for the area of a parallelogram is $A = \dfrac{h}{2}(a + b)$. Rewrite the formula with b as the subject.

_____

_____

_____

_____

8 The formula for converting degrees Celsius to Fahrenheit is $F = \dfrac{9}{5}C + 32$. Make C the subject of this formula.

_____

_____

_____

_____

9 The relationship between one side of a triangle (a) and cos A and the other two sides is $a^2 = b^2 + c^2 - 2bc \cos A$. Rewrite the formula with cos A as the subject.

_____

_____

_____

_____

10 The volume of a sphere is given by the formula $V = \dfrac{4}{3}\pi r^3$. Make r the subject of the formula.

_____

_____

_____

_____

Quadratic expressions

- An expression in which the highest power of the variable is **2** is known as a **quadratic** expression.
- Usually, these contain an x^2, but in factorised form, quadratic expressions may look like $(x \pm a)(x \pm b)$ or $x(x \pm a)$.
- When graphed, quadratics form curves known as **parabolas**:

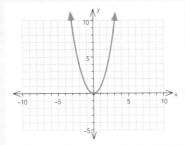   or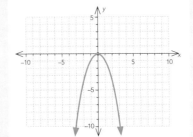

Expanding quadratic expressions

Remember **FOIL**: multiply the   **F**irsts
                            Outers
                            Inners
                            Lasts

F	O	I	L

e.g. $(x + 3)(x - 9) = x^2 - 9x + 3x - 27$
$\qquad\qquad\qquad\quad = x^2 - 6x - 27$

> Combine like terms.

Examples:

1 $(x + 4)(x + 7) = x^2 + 7x + 4x + 28$
$\qquad\qquad\qquad\quad = x^2 + 11x + 28$

3 $(x + 8)(x - 8) = x^2 - 8x + 8x - 64$
$\qquad\qquad\qquad\quad = x^2 - 64$

> An expression in this form is called 'the difference of two squares'.

2 $(x + 3)^2 = (x + 3)(x + 3)$
$\qquad\qquad\quad = x^2 + 3x + 3x + 9$
$\qquad\qquad\quad = x^2 + 6x + 9$

4 $(5 + x)(x - 2) = 5x - 10 + x^2 - 2x$
$\qquad\qquad\qquad\quad = x^2 + 3x - 10$

> While order of the terms doesn't matter, it's conventional to write your answer in this order.

28

Expand and simplify these.

1 $(x + 2)(x + 6)$

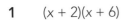

2 $(x + 7)(x - 1)$

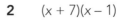

3 $(x - 3)(x + 4)$

4 $(x - 5)(x - 9)$

ISBN: 9780170477468

5 $(x + 6)^2$

_____

_____

6 $(3 + x)(x + 8)$

_____

_____

7 $(x - 5)^2$

_____

_____

8 $(x - 8)(x - 9)$

_____

_____

9 $(x + 4)(4 + x)$

_____

_____

10 $(x + 7)(x - 7)$

_____

_____

11 $(x + 11)^2$

_____

_____

12 $(x + 6)(-6 + x)$

_____

_____

13 $(8 + x)^2$

_____

_____

14 $(2 - x)(x + 7)$

_____

_____

15 $(3 + x)(4 - x)$

_____

_____

16 $(-1 - x)(10 + x)$

_____

_____

17 $(2 - x)^2$

_____

_____

18 $(9 - x)(5 - x)$

_____

_____

Factorising quadratic expressions

Steps:

1 List all the factors of the constant: $x^2 + 2x - 24$

1, 24
2, 12
3, 8
4, 6

2 Select the pair that could add or subtract to give the coefficient of x: $x^2 + 2x - 24$

−4 + 6 = 2

3 The factors are $(x - 4)(x + 6)$ or $(x + 6)(x - 4)$

Note: You can swap the order of the brackets but **not** the signs.

Check your answer by expanding the brackets using FOIL — you should get the original expression: $(x - 4)(x + 6) = x^2 + 2x - 24$

Examples:

1 $x^2 + 10x + 16 = (x + 2)(x + 8)$

1, 16
2, 8
4, 4

2 $x^2 - 8x + 15 = (x - 3)(x - 5)$

1, 15
−3, −5

3 $x^2 - 1x - 12 = (x + 3)(x - 4)$
or $(x - 4)(x + 3)$

1, −12
2, −6
3, −4

4 $x^2 - 81 = x^2 + 0x - 81 = (x + 9)(x - 9)$
or $(x - 9)(x + 9)$

Insert a 'fake' x term.

1, 81
9, −9

Factorise the following expressions.

1　$x^2 + 9x + 14$

_____

_____

2　$x^2 + 12x + 27$

_____

_____

3　$x^2 + 14x + 40$

_____

_____

4　$x^2 + 12x + 11$

_____

_____

5 $x^2 + 22x + 120$

_____

_____

6 $x^2 - 5x - 24$

_____

_____

7 $x^2 + 7x - 44$

_____

_____

8 $x^2 - 5x + 6$

_____

_____

9 $x^2 - 18x + 80$

_____

_____

10 $x^2 - x - 30$

_____

_____

11 $x^2 - 25$

_____

_____

12 $3 + 4x + x^2$

_____

_____

13 $15x + 26 + x^2$

_____

_____

14 $x^2 - 16$

_____

_____

15 $-4 + x^2$

_____

_____

16 $16x + x^2 + 63$

_____

_____

17 $9 - x^2$

_____

_____

18 $-7x + 6 + x^2$

_____

_____

Solving quadratic equations

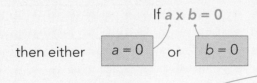

If $a \times b = 0$

then either $\boxed{a = 0}$ or $\boxed{b = 0}$

Remember, the times (x) sign has not been written in here.

So, if we have a quadratic equation $(x - 2)(x + 6) = 0$

then either $\boxed{(x - 2) = 0}$ or $\boxed{(x + 6) = 0}$

So either $x = 2$ or $x = -6$

Note:
- Most quadratic equations that you come across will have **two solutions**. However, some will have just one solution, and some will have no real solutions.
- In practical situations, there are usually two solutions, but only one will be practical.

Examples:

1 $(x + 9)(x - 1) = 0$

Either $(x + 9) = 0$

$x = -9$

or $(x - 1) = 0$

$x = 1$

2 $(x + 3)^2 = 0$

$\therefore (x + 3) = 0$

One solution only.

$x = -3$

3 $x(x - 10) = 0$

Either $x = 0$

or $(x - 10) = 0$

$x = 10$

4 $x - 4x - 5 = 0$

$(x - 5)(x + 1) = 0$

Either $(x - 5) = 0$

$x = 5$

Factorise first.

or $(x + 1) = 0$

$x = -1$

Solve the following equations.

1 $(x + 6)(x - 3) = 0$

_____

_____

2 $(x - 5)(x + 2) = 0$

_____

_____

3 $(x - 1)(x - 8) = 0$

_____

_____

4 $(x + 4)(x + 7) = 0$

_____

_____

5 $(x + 3)(x - 10) = 0$

_____

_____

6 $(x - 4)^2 = 0$

_____

_____

7 $(x + 7)^2 = 0$

_____

_____

8 $(x - 0.5)^2 = 0$

_____

_____

9 $(x + 1)(4 - x) = 0$

_____

_____

10 $(x - 2)(8 + x) = 0$

_____

_____

11 $x(x - 5) = 0$

_____

_____

12 $-x(x + 9) = 0$

_____

_____

13 $x^2 + 7x + 10 = 0$

_____

_____

_____

14 $x^2 - 2x = 3$

_____

_____

_____

15 $x^2 - 64 = 0$

_____

_____

_____

16 $2x - 80 + x^2 = 0$

_____

_____

_____

17 $x^2 - 13x - 68 = 0$

_____

_____

18 $x^2 + 182 = 27x$

_____

_____

Forming and solving quadratic equations

Hints:
- Read what the question asks for. Call this x.
- If there are two things to find, call the smaller one x and the bigger one $(x + ?)$.
- Usually you will get two answers. **Test both to see if they make sense.** If you have to reject one answer because it doesn't make sense, **explain** why.

1 Problems with shapes

Hint: If you are not given one, do your own sketch and add all the information to it.

Example: A rectangle has a height h and a base that is 3 cm longer than the height. Its area is 180 cm². Calculate its dimensions.

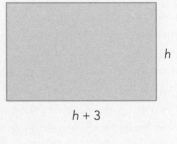

h

$h + 3$

Area of a rectangle = base x height = $(h + 3)$ x h

$\qquad\qquad\qquad\qquad\qquad = h(h + 3)$

$\therefore \quad h(h + 3) = 180$

$\qquad h^2 + 3h = 180$

$\qquad h^2 + 3h - 180 = 0$

$\qquad (h - 12)(h + 15) = 0$

$\qquad h = 12$ or -15

> h cannot be a negative value, so $h = 12$.

So the height of the rectangle is 12 cm and the base is $(12 + 3) = 15$ cm.

2 Problems with numbers

Finding unknown numbers:
- Even numbers: call these $2x$, e.g. 10, where $x = 5$.
- Odd numbers: call these $2x + 1$, e.g. 15, where $x = 7$.
- Consecutive numbers come after each other: call these x, $x + 1$, $x + 2$, e.g. 6, 7, 8, where $x = 6$.
- Consecutive even numbers: call these $2x$, $2x + 2$, $2x + 4$, e.g. 6, 8, 10, where $x = 3$.
- Consecutive odd numbers: call these $2x + 1$, $2x + 3$, $2x + 5$, e.g. 9, 11, 13, where $x = 4$.
- Age questions: give the youngest the age x.

Example: Ari thinks of a positive number. He subtracts four, squares the result, then adds one. His answer is 37. What number was he thinking of?

Call the number he first thought of x. $\qquad\qquad\qquad x$

He subtracts four, $\qquad\qquad\qquad\qquad\qquad\qquad\quad x - 4$

then squares the result, $\qquad\qquad\qquad\qquad\quad (x - 4)^2$

then adds one. $\qquad\qquad\qquad\qquad\qquad\qquad (x - 4)^2 + 1$

$(x - 4)^2 + 1 = 37$	or	$(x - 4)^2 + 1 = 37$
$x^2 - 8x + 16 + 1 = 37$		$(x - 4)^2 = 36$
$x^2 - 8x - 20 = 0$		$x - 4 = \pm 6$
$(x - 10)(x + 2) = 0$		$\therefore x = 10$ or $x = -2$
$\therefore x = 10$ or $x = -2$		

> When you have an expression like $(x - 4)^2 = 36$, and you take the square root of both sides, you must consider 2 answers: ± 6.

His number was positive, so -2 is not a possibility, **so he thought of 10.**

> Remember – always answer the question in a sentence.

 ISBN: 9780170477468

Write and solve quadratic equations to find the unknown numbers.

1 A rectangle has a height of $x - 1$ and a base of $x + 5$. Its total area is 55 m². Calculate the lengths of the sides.

_____

_____

_____

_____

$x - 1$

$x + 5$

2 Raukura thinks of a number (n). She squares it, then multiplies the answer by seven. Her answer is 448. What numbers could she have been thinking of?
Hint: There are two possible numbers.

_____

_____

_____

3 Claudia's dad is 28 years older than she is. Seven years ago, the product of their ages was 204. How old are they now?

_____

_____

_____

4 The base of a triangle is three times its vertical height. Its area is 216 cm². Calculate the length of the base and height.

_____

_____

_____

_____

h

$3h$

5 Wally had a bag of lollies. His sister had seven more than he has. They are both given another two lollies, and then the product of the number of lollies they each have is 170. How many lollies did Wally start with?

_____

_____

_____

6 The top and bottom of a rectangle are 16 cm longer than its sides. The area of the rectangle is 1380 mm². Calculate the length of one side of the rectangle.

_____

_____

_____

7 The sides of a right-angled triangle are x m, (x + 7) m and (x + 8) m. Use the Theorem of Pythagoras to calculate the lengths of the three sides.

_____

_____

_____

_____

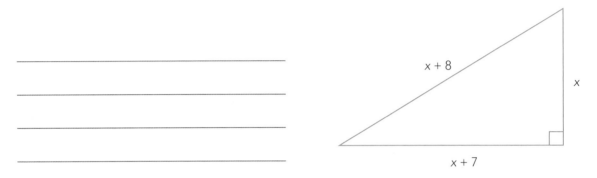

8 Gloria is the youngest of three siblings, each born two years apart. The product of the ages of the younger two siblings is 24 less than the product of the ages of the older two siblings. Calculate the ages of the three children.

_____

_____

_____

9 A rectangle is 8y cm wide and 5y cm high. A smaller rectangular shape is cut from this rectangle. This smaller rectangle is 3y cm wide and 2y cm high. The remaining area is 1224 cm². Calculate the value of y, and the dimensions of the big rectangle.

_____

_____

_____

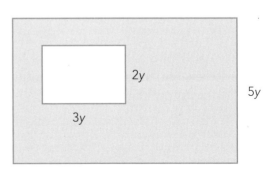

Plotting quadratic equations

Plotting quadratics:
- A number of programs will do this for you.
- However, if a hand sketch is required, tables are a useful tool.
- The turning point of a parabola is called the vertex.
- Note: The vertex is not pointed.

Example: Draw the graph of $y = x^2$.

x	y
3	9
2	4
1	1
0	0
−1	1
−2	4
−3	9

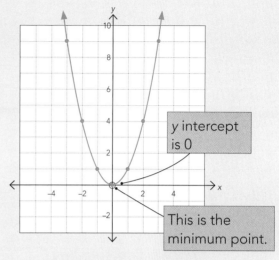

y intercept is 0

This is the minimum point.

All parabolas are this shape, but they can be shifted, turned upside down or stretched, or any combination of these. Here are some examples:

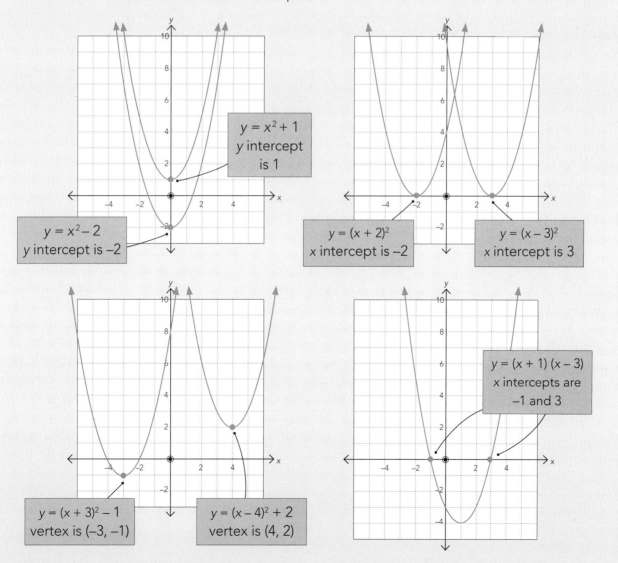

$y = x^2 + 1$
y intercept is 1

$y = x^2 - 2$
y intercept is −2

$y = (x + 2)^2$
x intercept is −2

$y = (x - 3)^2$
x intercept is 3

$y = (x + 3)^2 - 1$
vertex is (−3, −1)

$y = (x - 4)^2 + 2$
vertex is (4, 2)

$y = (x + 1)(x - 3)$
x intercepts are −1 and 3

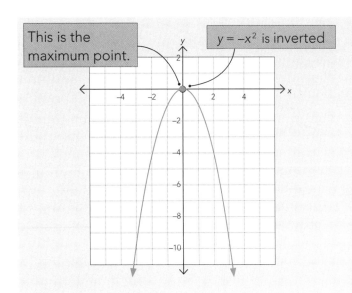

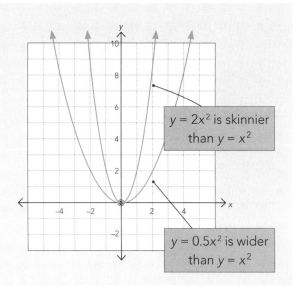

Complete the tables and draw the graphs for the following equations.

1 $y = x^2 + 2$

x	y
3	
2	
1	
0	
−1	
−2	
−3	

The vertex is a minimum/maximum at

(_____, _____).

2 $y = (x - 1)^2 + 2$

x	y
4	
3	
2	
1	
0	
−1	
−2	

The vertex is a minimum/maximum at

(_____, _____).

3 $y = -(x + 2)^2 + 8$

x	y
1	
0	
−1	
−2	
−3	
−4	
−5	

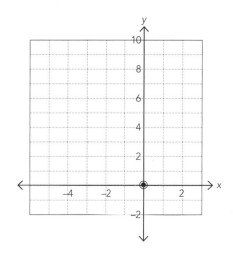

The vertex is a minimum/maximum at (_____, _____).

4 $y = -(x + 2)(x - 4)$

x	y
4	
3	
2	
1	
0	
−1	
−2	

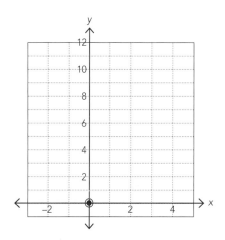

The vertex is a minimum/maximum at (_____, _____).

5 $y = 0.75x^2$

x	y
4	
2	
0	
−2	
−4	

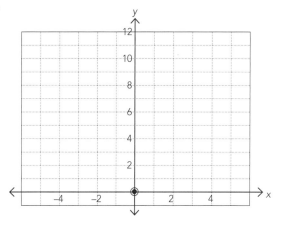

The vertex is a minimum/maximum at (_____, _____).

6 $y = 2(x - 1)^2 + 3$

x	y
3	
2	
1	
0	
−1	

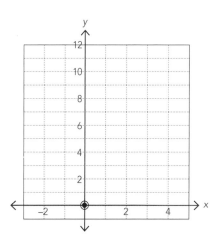

The vertex is a minimum/maximum at (_____, _____).

Optimising by graphing quadratics

- Optimising means finding the **maximum** or **minimum** values.
- The easiest way to do this is to give x a series of values. Then plot the relationship between x and the variable that needs to be optimised.
- Areas are **squared** relationships (measured in m^2, cm^2, etc.), so they are modelled by quadratic functions and parabolas.
- You could use technology to help with this.

Example: Henare has 38 m of netting which he will use to make a rectangular pen for his chickens. What are the dimensions for the pen which will maximise the area for the chickens?

Step 1:   The width and length of the pen must add to 19 m.
            So call the width x. Then the length is $(19 - x)$.
            $\therefore$ The area of the pen $(y) = x(19 - x)$.

Step 2:   Make a table showing the relationship between x and the area (y).

Width (x)	Length ($19 - x$)	Area: $y = x(19 - x)$
5	14	70
6	13	78
7	12	84
8	11	88
9	10	90
10	9	90
11	8	88

> The areas start to become smaller for $x > 10$, so there is no point in continuing the table.

Step 3:   Plot the points on a graph.

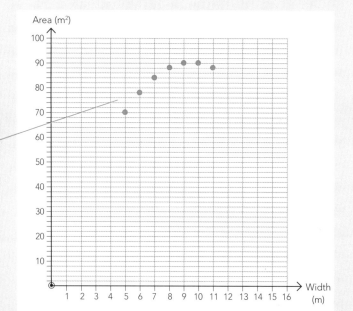

> The graph is a parabola, which is symmetrical $\therefore$ the maximum must occur halfway between the two equal top points, where $x = 9.5$.
> $x = 9.5 \Rightarrow$ Area $= 90.25$ m^2.
> The maximum is at $(9.5, 90.25)$.

Step 4:   Answer the question.
            The parabola is symmetrical, so pen which will maximise the area for Henare's chickens is a square with sides of 9.5 m, and its area will be 90.25 m^2.

     ISBN: 9780170477468

Answer the following questions.

1   Sam wants to make a triangular pen for his geese. He uses an existing fence for one side of the pen. He has a total length of 120 m of netting. The pen will be in the shape of a right-angled triangle.

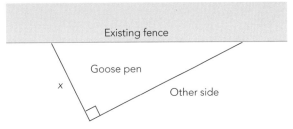

What are the lengths of the short sides of the triangle which result in a maximum area?

Complete the table to show the relationship between x, the other side of the goose pen and its area.

x	Other side $(120 - x)$	Area: $y = \dfrac{1}{2}x(120 - x)$
10		
20		

Plot the points on the graph.

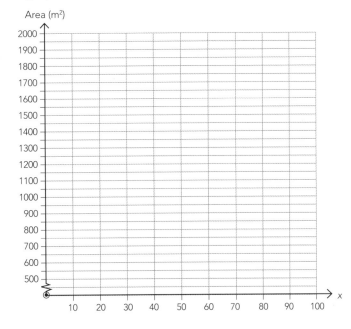

Answer the question. _____

_____

_____

2 Annie has 40 m of netting to make a rectangular duck enclosure. One side will consist of the hayshed wall, so she has only three sides to make from the netting.

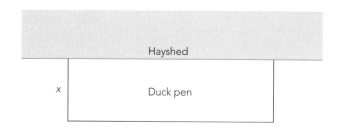

What are the dimensions which result in the maximum area inside the pen?

Complete the table to show the relationship between x, another side of the duck pen and its area.

Width (x)	Length (40 – 2x)	Area: y = x(40 – 2x)
6		

Plot the points on the graph.

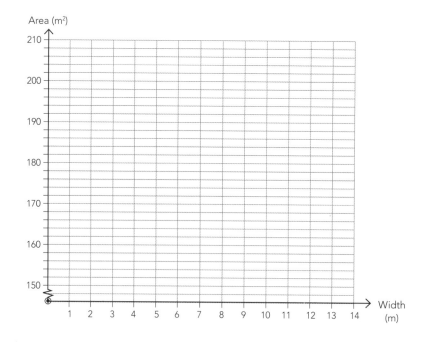

Answer the question. _____

_____

_____

 ISBN: 9780170477468

In some questions, you may be given the formula for the variable that is to be maximised.

3 A sky rocket is fired from the ground and its flight is described by the formula $y = 10x(9 - x)$, where y represents the height (m) of the rocket above the ground and x represents the time (s) since the rocket was fired.

How long did the rocket take to reach its maximum height, and what was its maximum height above the ground?

x	Height: $y = 10x(9 - x)$
1	

Plot the points on the graph.

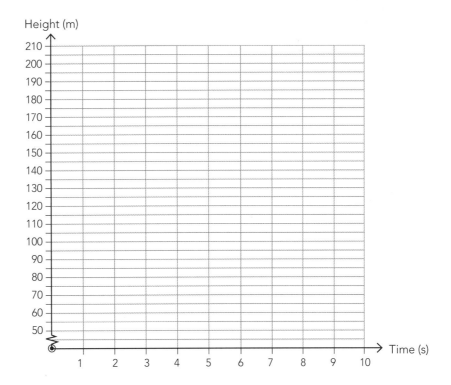

Answer the question. _____

_____

_____

Discuss any assumptions you have made.

_____

_____

The same method works for finding a minimum.

4 On a confidence course, a rope is hung from a horizontal bar which is 2.6 m above ground level.

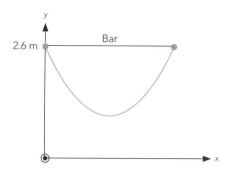

The shape formed by the rope approximates a parabola with the equation $y = 0.4(x - 2)^2 + 1$, where y represents the height of the rope above the ground and x represents the distance from the left end of the rope.

How high above the ground is the lowest point on the rope and how wide is the bar?
Hint: Plot x values at 0.5 m intervals.

x	Height: $y = 0.4(x - 2)^2 + 1$
0.5	
1	

Plot the points on the graph.

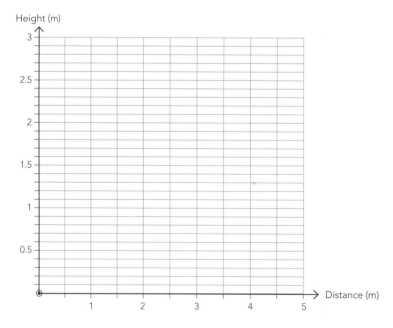

Answer the question. _____

_____

_____

 # Geometry and space

Fundamentals

Words

Write a definition for each of these terms.

Term	Description	Diagram
Acute angle	An angle that is less than 90°	
Obtuse angle		
Right angle		
Reflex angle		
Scalene triangle		
Isosceles triangle		
Equilateral triangle		
Right-angled triangle		
Supplementary angles		
Complementary angles		

33

Relationships

Angles

Relationship	Reason
$a + b = 180°$	Angles on a line add to 180°. (∠s on a line = 180°)
$a + b + c = 360°$	Angles at a point add to 360°. (∠s at a point = 360°)
$a = b$	Vertically opposite angles are equal. (vert opp ∠s =)
$a + b + c = 180°$	Angles in a triangle add to 180°. (∠s in Δ = 180°)
$a + b = c$	The exterior angle of a triangle = the sum of the interior opposite angles. (ext ∠ of Δ = sum of int opp ∠s)

 ISBN: 9780170477468

34

Parallel lines

Relationship	Reason
a = b	Alternate angles on parallel lines are equal. (These form a 'Z'.) (alt ∠s =, // lines)
a = b	Corresponding angles on parallel lines are equal. (These form an 'F'.) (corr ∠s =, // lines)
a + b = 180°	Co-interior angles on parallel lines add to 180°. (These form a 'C'.) (co-int ∠s add to 180°, // lines)

35

Polygons

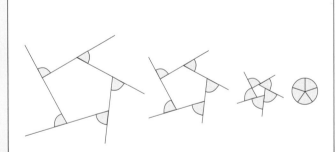	Exterior angles of a polygon add to 360°. (ext ∠s of polygon = 360°)
Example: Hexagon 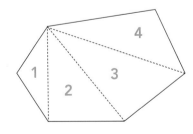 Interior angles add to 4 x 180° = 720°.	Interior angles of a polygon add to $(n - 2) \times 180°$. Each interior angle of a **regular** polygon $$= \frac{(n - 2) \times 180°}{n} = 180° - \frac{360°}{n}$$

Bearings

- Angles are always measured **clockwise** from north.
- Angles **must** have **three** digits.

Examples:

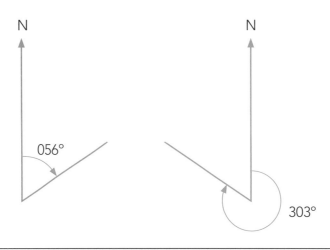

Compass directions

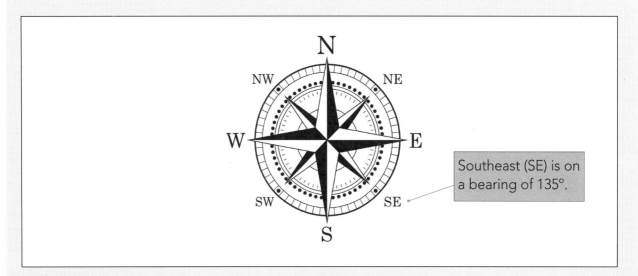

Southeast (SE) is on a bearing of 135°.

Angle of elevation and depression

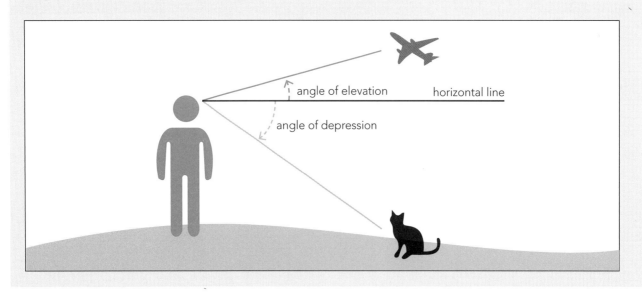

Similar triangles

- Similar triangles are the same shape because they are **enlargements** of each other.
- They have the same-sized angles as each other.

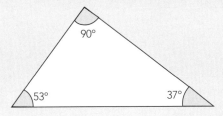
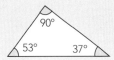

- However, they are not always drawn the same way up.

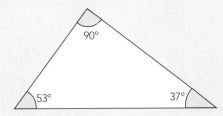
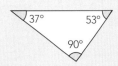

- The angles can also be in reverse order.

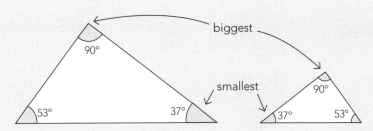

- The sides are also in proportion.

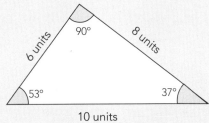

In this case: $\dfrac{\text{big triangle}}{\text{small triangle}} = \dfrac{10}{5} = \dfrac{8}{4} = \dfrac{6}{3} = 2$

This is called the scale factor.

- It is customary to label the vertices in corresponding order.

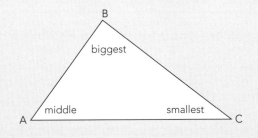
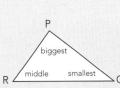

In this case: $\triangle$**BAC** is similar to $\triangle$**PRQ**.

Both are in **biggest, middle, smallest** order.

Justification of similar triangles using angles

- You need to show that **all** corresponding angles are equal.

Example:

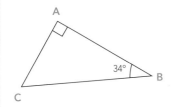 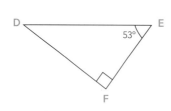 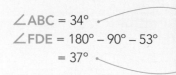

$\angle ABC = 34°$

$\angle FDE = 180° - 90° - 53°$

$\qquad = 37°$

The smallest angles are not the same, so these triangles are **not** similar.

Show whether these shapes are similar or not.

1

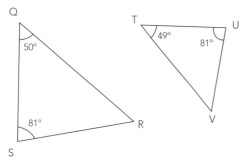

_____
_____
_____
_____

2

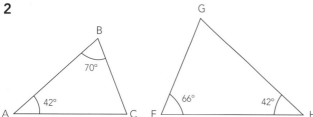

_____
_____
_____
_____
_____

3

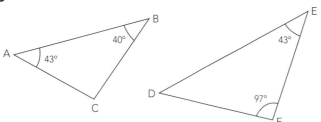

_____
_____
_____
_____
_____

4

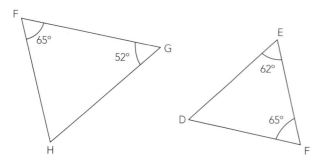

_____
_____
_____
_____

Justification of similar triangles using sides

- You need to show that **all** pairs of equivalent sides are proportional.

Example:

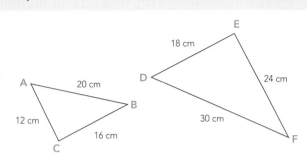

shortest side of small triangle
shortest side of big triangle

$$\frac{AC}{DE} = \frac{12}{18} = \frac{2}{3}$$

$$\frac{CB}{EF} = \frac{16}{24} = \frac{2}{3}$$

$$\frac{AB}{DF} = \frac{20}{30} = \frac{2}{3}$$

longest side of small triangle
longest side of big triangle

These triangles are similar because all their sides are proportional.

Show whether these triangles are similar or not.

1

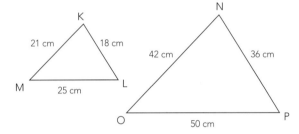

_____
_____
_____
_____

2

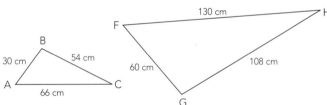

_____
_____
_____
_____

3

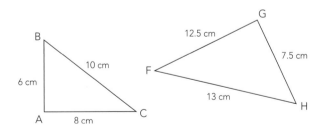

_____
_____
_____
_____

4

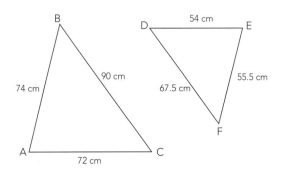

_____
_____
_____
_____

ISBN: 9780170477468

Calculating lengths of unknown sides

- Because the sides of similar triangles are in proportion, if we know the lengths of a pair of corresponding sides, we can use their proportion to calculate other sides.

Example:

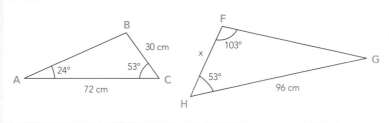

$\triangle ABC$ is similar to $\triangle GFH$

So $\dfrac{AC}{GH} = \dfrac{BC}{FH}$

$\dfrac{72}{96} = \dfrac{30}{x}$

$x = \dfrac{96 \times 30}{72}$

$x = 40$ cm

Find the missing lengths for these similar shapes.

1

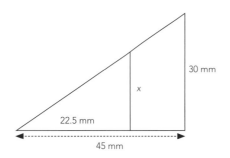

_____
_____
_____
_____

2

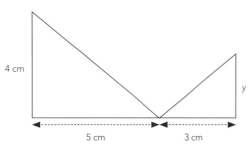

_____
_____
_____
_____
_____

3

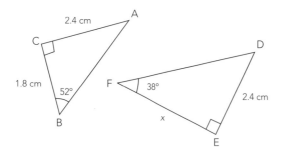

_____
_____
_____
_____
_____

4   Calculate the lengths of AB and AE.

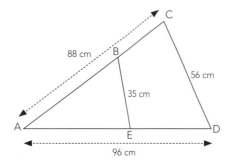

_____
_____
_____
_____

5 Anahera wants to know the height of this tree.
She measures the length of the tree's shadow
and then measures the length of her own
shadow. She is 165 cm tall. Calculate the
height of the tree.

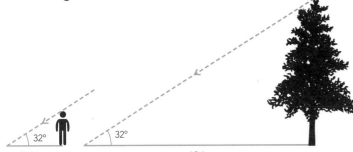

_____

_____

_____

_____

_____

_____

Discuss any assumptions that you made in your answer.

_____

_____

_____

6 The PTA is building a slide for the playground.
The two shaded posts need to be set 80 cm
into the ground. Calculate the total length of
timber needed for these two posts.

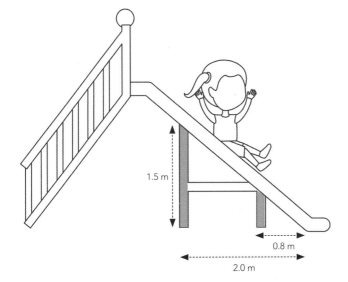

_____

_____

_____

_____

_____

_____

_____

_____

Discuss any assumptions that you made in your answer.

_____

_____

_____

Theorem of Pythagoras

- The Theorem of Pythagoras applies to **right-angled** triangles only.
- The longest side is the **hypotenuse**.
- The hypotenuse is always **opposite** the right angle.
- The theorem is used for finding **lengths** of sides.

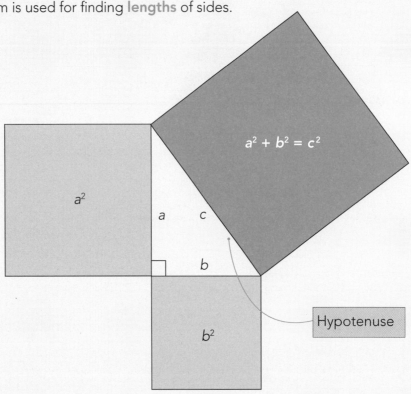

a^2

a c

$a^2 + b^2 = c^2$

b

b^2

Hypotenuse

short side² + short side² = hypotenuse²

$$a^2 \quad + \quad b^2 \quad = \quad c^2$$

Examples:

1 Finding the length of the hypotenuse
Calculate the length of c.

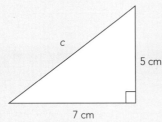

c

5 cm

7 cm

$$5^2 + 7^2 = c^2$$
$$\sqrt{5^2 + 7^2} = c$$
$$c = \sqrt{74}$$
$$c = 8.60 \text{ cm (2 dp)}$$

8.6 cm is a reasonable answer because it's longer than the other two sides.

2 Finding the length of short sides
Calculate the length of x.

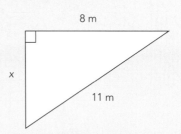

8 m

x

11 m

$$x^2 = 11^2 - 8^2$$
$$x = \sqrt{11^2 - 8^2}$$
$$x = \sqrt{57}$$
$$x = 7.55 \text{ m (2 dp)}$$

7.55 m is a reasonable answer because it's shorter than the hypotenuse.

    ISBN: 9780170477468

Calculate the unknown lengths of each triangle.

1

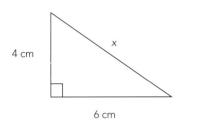

4 cm

x

6 cm

_____

_____

_____

2

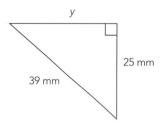

y

25 mm

39 mm

_____

_____

_____

3

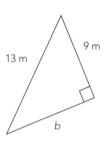

9 m

13 m

b

_____

_____

_____

4

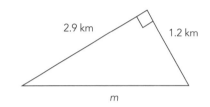

2.9 km

1.2 km

m

_____

_____

_____

5

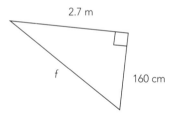

2.7 m

f

160 cm

_____

_____

_____

6

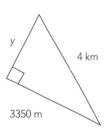

y

4 km

3350 m

_____

_____

_____

7 This equilateral triangle has sides of 7 cm.

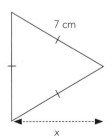

7 cm

x

_____

_____

_____

8 This isosceles triangle has a vertical height of 43 m.

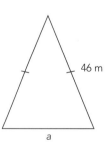

46 m

a

_____

_____

_____

Multi-step and 2D practical problems

Example: Calculate the length of CD.

Step 1: AC is common to both triangles, so calculate its length.

$$AC^2 = AB^2 + BC^2$$
$$AC = \sqrt{15^2 + 8^2}$$
$$AC = 17 \text{ cm}$$

Step 2: Use the length of AC to calculate the length of CD.

$$AD^2 + CD^2 = AC^2$$
$$CD^2 = AC^2 - AD^2$$
$$CD = \sqrt{17^2 - 13^2}$$
$$CD = 10.95 \text{ cm (2 dp)}$$

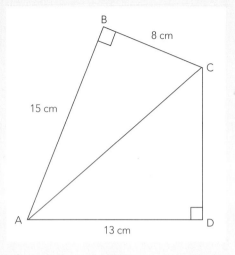

Answer the following questions.

1 Calculate the length of CD.

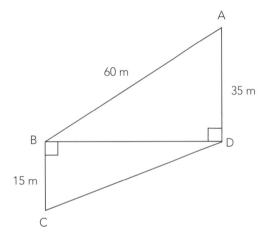

_____

_____

_____

_____

_____

_____

_____

2 Calculate the length of x.

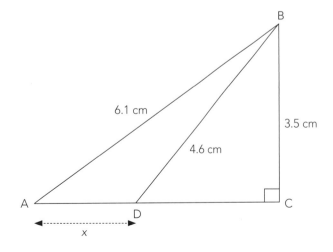

_____

_____

_____

_____

_____

_____

_____

ISBN: 9780170477468

3 Calculate the length *y*.

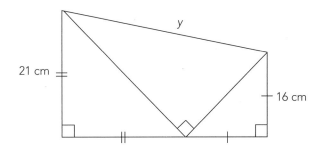

_____

_____

_____

_____

_____

4 Calculate the perimeter of the quadrilateral ABCD.

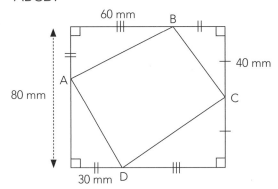

_____

_____

_____

_____

_____

_____

_____

5 The diagram shows a regular hexagon, with centre A. Calculate its perimeter.

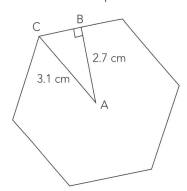

_____

_____

_____

_____

_____

_____

6 Calculate the length of CD.

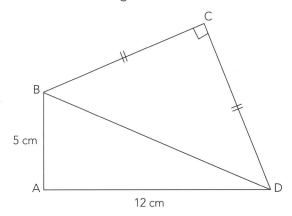

_____

_____

_____

_____

_____

_____

_____

7 Gia started her walk heading north for 6.3 km. Then she turned 270° clockwise and continued another 4.1 km. She then returned to her starting point. Sketch a diagram of her walk and calculate how far she walked in total.

_____

_____

_____

_____

8 People are stuck on the tenth floor of a building where the windows are 30 m above ground level. The maximum length for a fire engine's ladder is 32 m and the base of the ladder is located 2 m above ground level. Unfortunately, the engine can't get closer than 15 m from the base of the building. Will the ladder be long enough to reach the people? Support your answer with calculations.

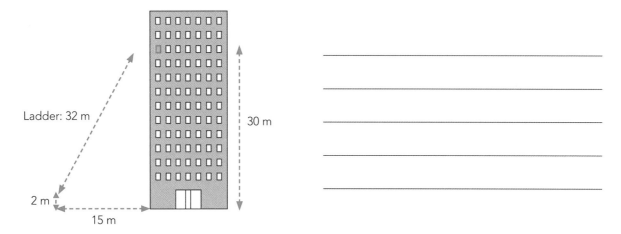

_____

_____

_____

_____

_____

9 Ngaere's grandmother stitched this design onto fabric. Before she frames it, she has to check whether it needs to be stretched to make the design exactly square. She carefully measures the distances shown. Use these measurements to help you decide whether the pattern is exactly square.

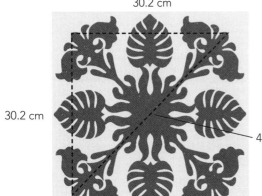

30.2 cm

30.2 cm

43.0 cm

_____

_____

_____

_____

_____

3D problems

- When asked to find a length or angle within a 3D shape, it is a good idea to draw the 2D shape which contains the length or angle.
- These usually involve right-angled triangles, so you can expect to use the Theorem of Pythagoras and trigonometry.

Example: This figure represents a cuboid.

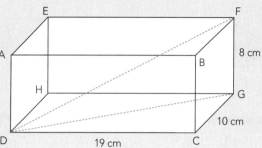

1 Calculate the length of DG.
DG lies in a triangle on the 'floor' of the cuboid:

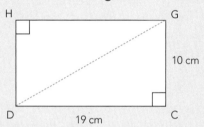

$DG^2 = 19^2 + 10^2$ (Pythagoras)
$\therefore DG = 21.471$ cm

2 Calculate the length of DF.
DF lies on the triangle DFG:

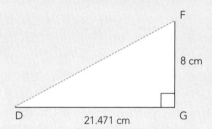

$DF^2 = 21.471^2 + 8^2$ (Pythagoras)
$\therefore DF = 22.913$ cm

Answer the following questions.

1 a This figure represents a cuboid.
Calculate the length of FH.

_____

_____

_____

b Calculate the length of AH.

_____

_____

c Calculate the length of BH.

_____

_____

2 This figure represents a right-angled triangular prism.
Calculate the length of AD.

_____

_____

_____

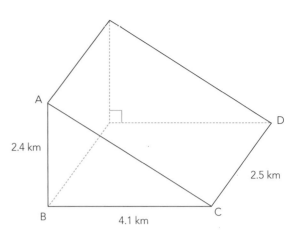

3 The green figure is the corner of a cube. The cube has edges of 25 cm and E is the centre of its base.
Calculate the length of AE.

_____

_____

_____

_____

_____

4 A cube has sides of 11 cm. Calculate the length of the green diagonal.

_____

_____

_____

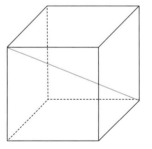

5 The base of this pyramid is a regular hexagon. Find the height of the pyramid.
Hint: You will need to use some polygon and triangle geometry.

_____

_____

_____

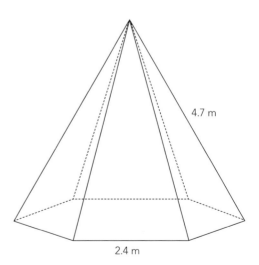

Trigonometry

- Trigonometry can also be used for calculations involving **right-angled triangles**.
- However, unlike calculations using the Theorem of Pythagoras, an **angle must** be involved.

Before you begin a calculation involving trigonometry, **label** the sides of the triangle:

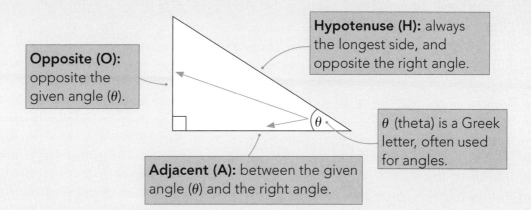

Opposite (O): opposite the given angle (θ).

Hypotenuse (H): always the longest side, and opposite the right angle.

Adjacent (A): between the given angle (θ) and the right angle.

θ (theta) is a Greek letter, often used for angles.

SOH CAH TOA

You will need to know about three trigonometric functions: **sin** θ (sine)
cos θ (cosine)
tan θ (tangent).

These are the rules in trigonometry:

$$\sin \theta = \frac{O}{H} \qquad \cos \theta = \frac{A}{H} \qquad \tan \theta = \frac{O}{A}$$

These are usually remembered as the 'word' **SOH CAH TOA**.

Organising **SOH CAH TOA** into triangles can help you to work out how to use it:

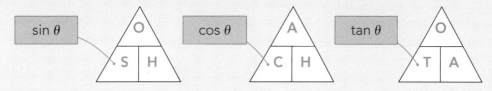

Process

Step 1: **Label** the sides that are involved in the problem with A, O and H.

Step 2: Decide **which relationship** involves these sides.

Step 3: **Substitute** the values from the triangle.

Step 4: **Think about your answer — does it seem sensible?**

Finding sides

- We can find an unknown side of a right-angled triangle using trigonometry if we have an angle (other than the right angle) and another side.

Using sine
Examples:

1 Finding a **short** side (O) using **sine**.

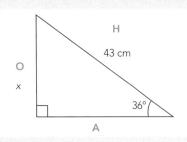

$$\sin \theta = \frac{O}{H}$$

$$\sin 36° = \frac{x}{43}$$

$$x = 43 \times \sin 36°$$

$$= 25.27 \text{ cm (2 dp)}$$

25.27 cm seems a reasonable answer because it's shorter than the hypotenuse.

2 Finding a **long** side (H) using **sine**.

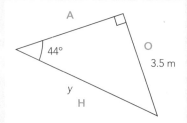

$$\sin \theta = \frac{O}{H}$$

$$\sin 44° = \frac{3.5}{y}$$

$$y \times \sin 44° = 3.5$$

$$y = \frac{3.5}{\sin 44°}$$

$$y = 5.04 \text{ m (2 dp)}$$

5.04 m seems a reasonable answer because it's the hypotenuse and it's longer than the opposite side.

Use sine to calculate the unknown length in each triangle. Give your answers to 2 dp.

1

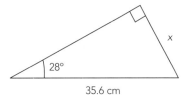

_____

_____

_____

2

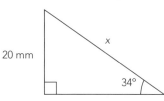

_____

_____

_____

3

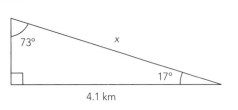

_____

_____

_____

4

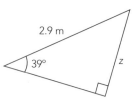

_____

_____

_____

 ISBN: 9780170477468

Using cosine

Examples:

1 Finding a **short** side (A) using **cosine**.

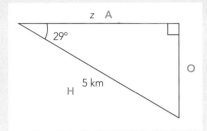

$$\cos \theta = \frac{A}{H}$$

$$\cos 29° = \frac{z}{5}$$

$$z = 5 \times \cos 29°$$

$$= 4.37 \text{ km (2 dp)}$$

4.37 km seems a reasonable answer because it's shorter than the hypotenuse.

2 Finding a **long** side (H) using **cosine**.

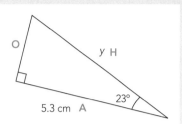

$$\cos \theta = \frac{A}{H}$$

$$\cos 23° = \frac{5.3}{y}$$

$$y \times \cos 23° = 5.3$$

$$y = \frac{5.3}{\cos 23°}$$

$$y = 5.76 \text{ cm (2 dp)}$$

5.76 cm seems a reasonable answer because it's the hypotenuse and it's longer than the adjacent side.

43

Use cosine to calculate the unknown length in each triangle. Give your answers to 2 dp.

5

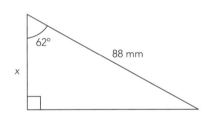

_____

_____

_____

6

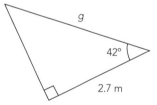

_____

_____

_____

7

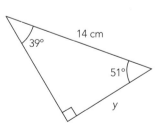

_____

_____

_____

8

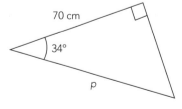

_____

_____

_____

Using tangent
Examples:

1 Finding an **opposite** side (O) using **tangent**.

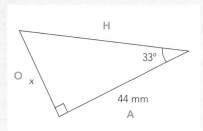

$$\tan \theta = \frac{O}{A}$$

$$\tan 33° = \frac{x}{44}$$

$$x = 44 \times \tan 33°$$

$$= 28.57 \text{ mm (2 dp)}$$

28.57 mm is a reasonable answer because the shortest side must be opposite the smallest angle.

2 Finding an **adjacent** side (A) using **tangent**.

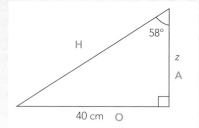

$$\tan \theta = \frac{O}{A}$$

$$\tan 58° = \frac{40}{z}$$

$$z \times \tan 58° = 40$$

$$z = \frac{40}{\tan 58°}$$

$$z = 24.99 \text{ cm (2 dp)}$$

24.99 cm is a reasonable answer because the shortest side must be opposite the smallest angle.

Use tangent to calculate the unknown length in each triangle. Give your answers to 2 dp.

9

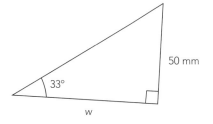

_____
_____
_____

10

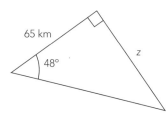

_____
_____
_____

11

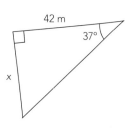

_____
_____
_____

12

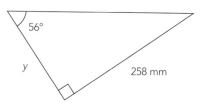

_____
_____
_____

Mixing it up

Calculate the unknown length in each triangle. Give your answers to 2 dp.

1

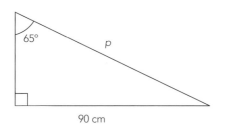

65°

p

90 cm

_____

_____

_____

2

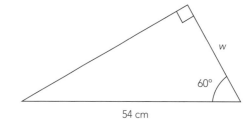

w

60°

54 cm

_____

_____

_____

3

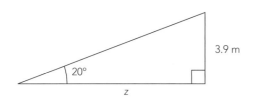

3.9 m

20°

z

_____

_____

_____

4

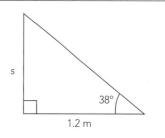

s

38°

1.2 m

_____

_____

_____

5

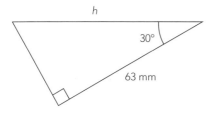

h

30°

63 mm

_____

_____

_____

6

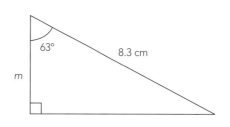

63°

8.3 cm

m

_____

_____

_____

7

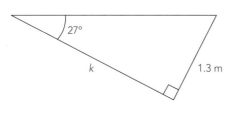

27°

k

1.3 m

_____

_____

_____

8

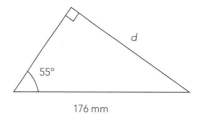

d

55°

176 mm

_____

_____

_____

Multi-step problems

Calculate the coordinates of the point (x, y).

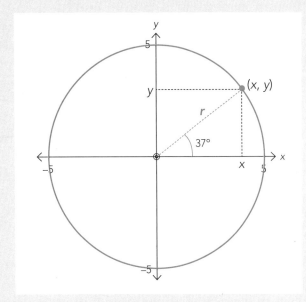

To find x: $\cos 37° = \dfrac{A}{H} = \dfrac{x}{r}$

$= \dfrac{x}{5}$

$x = 5 \times \cos 37°$

$= 4.0 \ (1 \ dp)$

To find y: $\sin 37° = \dfrac{O}{A} = \dfrac{y}{r}$

$= \dfrac{y}{5}$

$y = 5 \times \sin 37°$

$= 3.0 \ (1 \ dp)$

So the coordinates of (x, y) are $(4, 3)$.

Calculate the unknown length in each triangle. Give your answers to 2 dp.

1 Hint: Bisect the triangle.

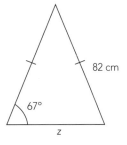

82 cm

67°

z

_____

_____

_____

2

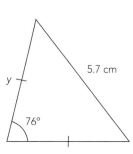

5.7 cm

y

76°

_____

_____

_____

3

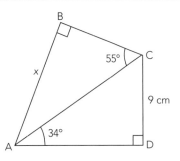

B

55°

C

x

9 cm

34°

A   D

_____

_____

_____

4

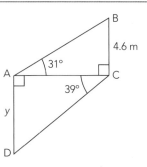

B

4.6 m

31°

A   C

39°

y

D

_____

_____

_____

5 Calculate the length of AD.

_____

_____

_____

_____

_____

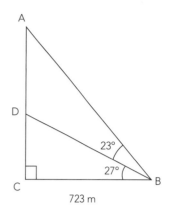

6 Use trigonometry to calculate the length of *y*.

_____

_____

_____

_____

_____

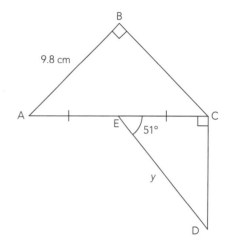

7 Calculate the length of DC.

_____

_____

_____

_____

_____

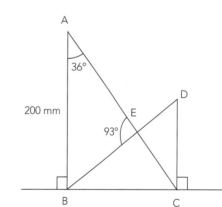

8 Calculate the coordinates (*x*, *y*).

_____

_____

_____

_____

_____

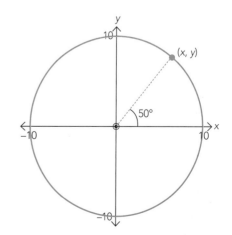

Finding angles

- We can find an unknown angle of a right-angled triangle using trigonometry if we know the **lengths of two sides**.

Examples:
Using sine

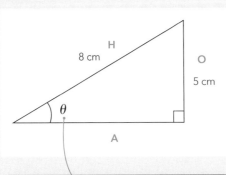

$$\sin \theta = \frac{O}{H}$$

$$\sin \theta = \frac{5}{8}$$

'undo' sin

$$\theta = \sin^{-1}\left(\frac{5}{8}\right)$$

'undo' sin

$$\theta = 38.7° \text{ (1 dp)}$$

Use the 'inverse sin' button on your calculator.

Put brackets around the fraction.

θ (theta) is a symbol that is often used for an unknown angle.

Think about your answer — does it seem reasonable? In this case, 38.7° is reasonable because it is less than 90°, and the smallest angle is opposite the shortest side.

Using cosine

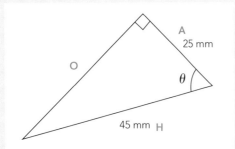

$$\cos \theta = \frac{A}{H}$$

$$\cos \theta = \frac{25}{45}$$

'undo' cos

$$\theta = \cos^{-1}\left(\frac{25}{45}\right)$$

'undo' cos

$$\theta = 56.3° \text{ (1 dp)}$$

Think about your answer. θ is opposite the longer of the two perpendicular sides, so an answer between 45° and 90° is expected.

Using tangent

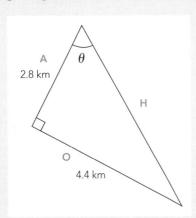

$$\tan \theta = \frac{O}{A}$$

$$\tan \theta = \frac{4.4}{2.8}$$

'undo' tan

$$\theta = \tan^{-1}\left(\frac{4.4}{2.8}\right)$$

'undo' tan

$$\theta = 57.5° \text{ (1 dp)}$$

Think about your answer. θ is opposite the longer of the two perpendicular sides, so an answer between 45° and 90° is expected.

Calculate the unknown angle of each triangle. Give your answers to 1 dp.

1

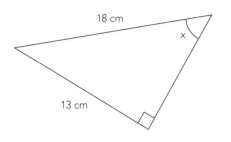

18 cm

x

13 cm

_____

_____

_____

2

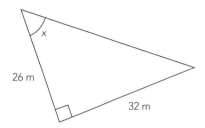

x

26 m

32 m

_____

_____

_____

3

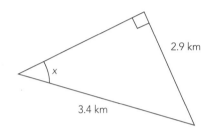

2.9 km

x

3.4 km

_____

_____

_____

4

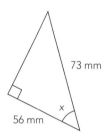

73 mm

x

56 mm

_____

_____

_____

5

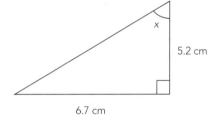

x

5.2 cm

6.7 cm

_____

_____

_____

6

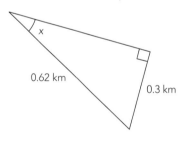

x

0.62 km

0.3 km

_____

_____

_____

7

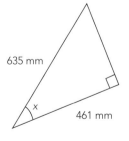

635 mm

x

461 mm

_____

_____

_____

8

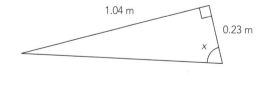

1.04 m

0.23 m

x

_____

_____

_____

Putting it together

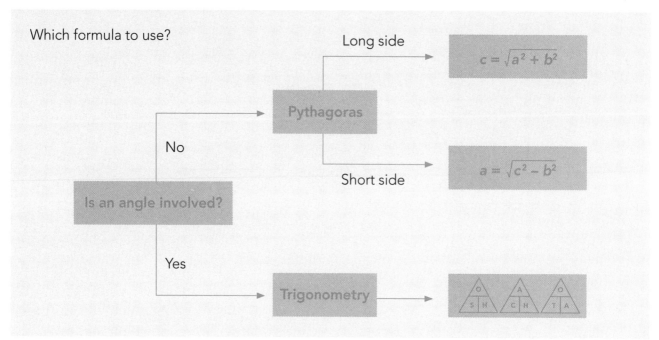

Which formula to use?

Pythagoras

Long side → $c = \sqrt{a^2 + b^2}$

Short side → $a = \sqrt{c^2 - b^2}$

Is an angle involved?

No

Yes

Trigonometry → SOH CAH TOA

Calculate the unknown lengths or angles of the triangles. Round lengths to 2 dp and angles to 1 dp. 48

1

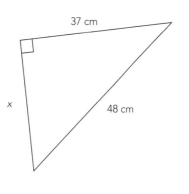

37 cm

x

48 cm

_____

_____

_____

2

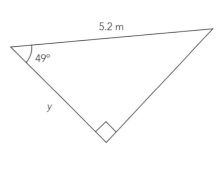

5.2 m

49°

y

_____

_____

_____

3

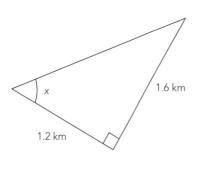

x

1.6 km

1.2 km

_____

_____

_____

4

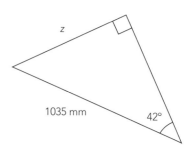

z

1035 mm

42°

_____

_____

_____

ISBN: 9780170477468

5

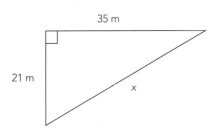

35 m

21 m

x

_____

_____

_____

6

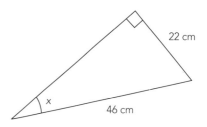

22 cm

x

46 cm

_____

_____

_____

7

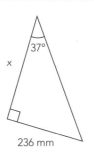

37°

x

236 mm

_____

_____

_____

8

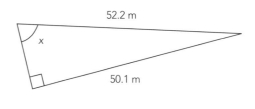

52.2 m

x

50.1 m

_____

_____

_____

9 Calculate x and y.

_____

_____

_____

_____

_____

_____

_____

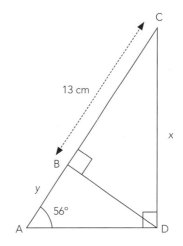

C

13 cm

x

B

y

56°

A   D

10 Calculate x and y.

_____

_____

_____

_____

_____

_____

_____

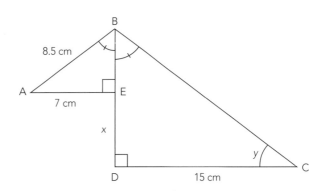

B

8.5 cm

A

7 cm   E

x

D   15 cm   C

y

Applications

Answer the following questions.

1   A boat has its anchor at point P at a depth of 18 m. When at full stretch, the anchor line is at 21° with the seabed. In order for the boat to be held securely, the anchor line must be at least three times the depth of the water. Is the anchor line long enough?

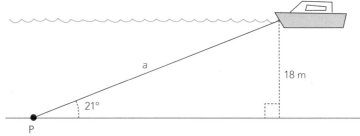

2   Alex is standing 90 m from a tree. The angle of elevation from his eyes to the top of the tree is 28°. His eyes are 1.58 m above ground level. How high is the tree?

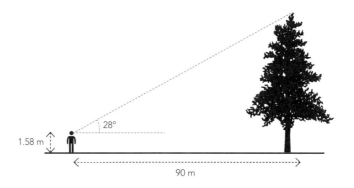

3   The shadow cast by a 7 m flag pole is 11 m long. Calculate the angle (θ) that the sun's rays make with the ground.

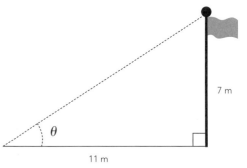

4   Heremia is standing 100 m away from a tree. The angle of elevation is 26°. Heremia is 175 cm tall. How tall is the tree?

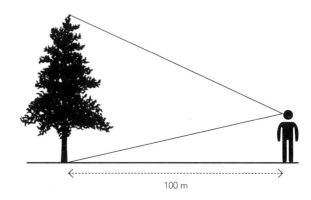

5   A wheelchair ramp is being built for a shop. The angle between the ramp and the pavement must be no more than 4.8°. The floor of the shop is 31 cm higher than the pavement. If the maximum horizontal distance for the ramp is 4 m, will the ramp comply with the building code?

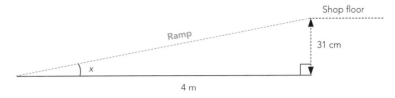

_____

_____

_____

6   Henare is standing on a cliff edge looking at a yacht. From his eye level, he measures the angles to the yacht and a rock which he knows is 37 m from the base of the cliff. Calculate the yacht's distance from the cliff.

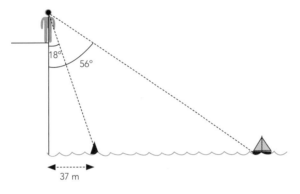

_____

_____

_____

7   Tane would like to know the height of the head (h) on this Rapa Nui statue. He measured the distance from the chin to the ground and found it was 2.20 m. He moved a distance away and measured the angles of elevation to the chin and the top of the head.
Find the overall height of the statue.

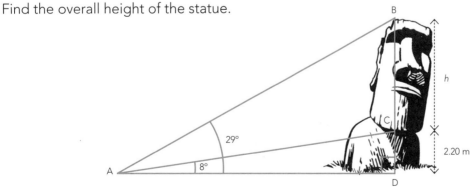

_____

_____

_____

Bearings

- Bearings are used in navigation to define direction in a horizontal plane.
- Direction can be given in terms of north, south, east and west.
- Bearings are measured in degrees in a **clockwise** direction **from north (000° or 360°)**.
- All bearings must have **three digits**.

Directions	**Bearings**
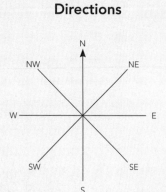	

Example: The first two stages of the school cross-country course (ABC) are shown on the diagram. The first stage is 350 m and on a bearing of 126°. The second is 500 m and on a bearing of 036°.

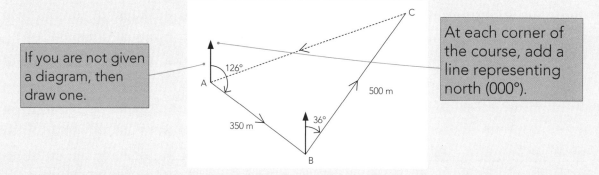

> If you are not given a diagram, then draw one.

> At each corner of the course, add a line representing north (000°).

a Calculate the length of the final stage (CA).

Calculate ∠ABC:

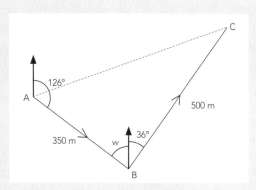

$w = 54°$ (co-int ∠s add to 180°, // lines)
∴ ∠ABC = 90°

∴ Theorem of Pythagoras can be used to calculate the length of AC:

$CA = \sqrt{350^2 + 500^2} = 610.3$ m (1 dp)

b Calculate the bearing of the final stage.

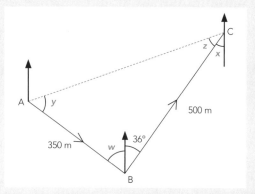

$x = 36°$ (alt ∠s =, // lines)

$y = \tan^{-1} \dfrac{500}{350} = 55.0°$

$z = 35°$ (∠ sum of Δ = 180°)

∴ Bearing = 180° + 36° + 35°

= 251°

1 The first two stages of another cross-country course (ABC) are shown on the diagram. The first stage is 680 m and on a bearing of 105°. The second is 300 m and on a bearing of 195°.

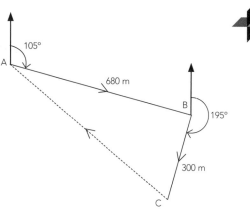

a Calculate the length of the final stage (CA).

_____

_____

_____

_____

_____

b Calculate the bearing of the final stage.

_____

_____

_____

_____

_____

2 A yacht sails 3.37 nautical miles (NM) from point A on a bearing of 330°. It changes course at B and sails 7.41 nautical miles on a bearing of 060° to reach point C.

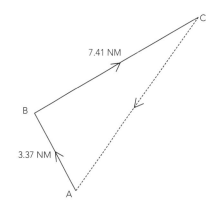

a Calculate how far point C is from their starting point (A).

_____

_____

_____

_____

b Calculate the bearing of a course that takes them from C back to their starting point (A).

_____

_____

_____

_____

c Discuss any assumptions you have made in your answers.

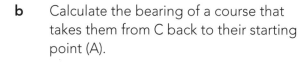

_____

_____

Measurement

Unit conversion

Units of length, mass and capacity

Use the following charts to help you convert units.

Length

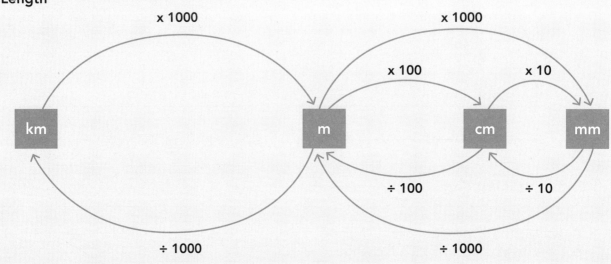

Mass

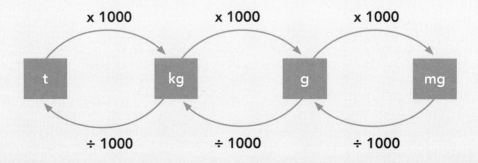

Capacity (fluids)

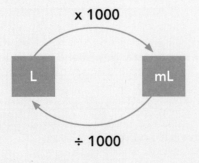

Units of area

- Length is measured in mm, cm, m or km.
- Area is measured in mm², cm², m², ha (hectares) or km².

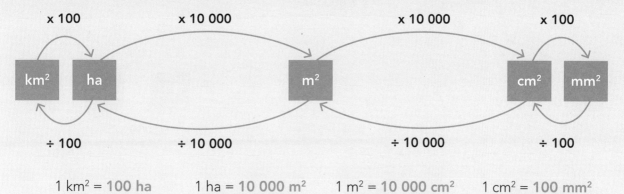

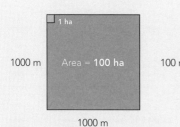 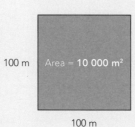 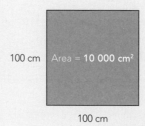 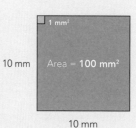

Example: Convert 24 950 cm² into m².

$$1 \text{ cm}^2 = 0.0001 \text{ m}^2$$
$$\therefore 24\,950 \text{ cm}^2 = 24\,950 \text{ cm}^2 \div 10\,000$$
$$= 2.495 \text{ m}^2$$

> Your answer will be a smaller number, so **divide**.

Convert the following.

1 4 ha = _____

= _____ m²

2 26 400 cm² = _____

= _____ m²

3 0.19 km² = _____

= _____ ha

4 6 830 mm² = _____

= _____ cm²

5 1.97 m² = _____

= _____ mm²

6 0.0093 km² = _____

= _____ m²

7 640 000 cm² = _____

= _____ ha

8 0.34 ha = _____

= _____ m²

9 A house is 1530 cm long and 950 cm wide. Calculate its area in m².

_____

_____

10 The land the house sits on is 0.135 ha. If it is 45 m long, how wide is it?

_____

_____

Units of volume

- Volume is measured in mm^3, cm^3, mL, L or m^3.
- cm^3 are often used in medical contexts, but they are called '**cc**s' (**c**ubic **c**entimetres).

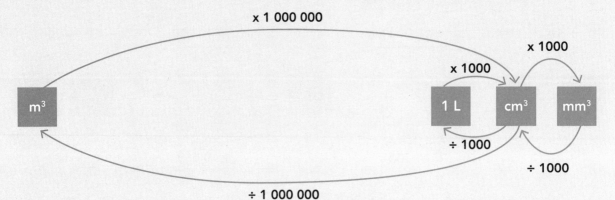

$1 cm^3 = 1 mL = 1000 mm^3$ $1 L = 1000 mL = 1000 cm^3$ $1 m^3 = 1\,000\,000\,cm^3$
$= 1000\,L$

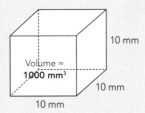 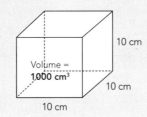 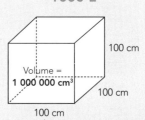

Example:
Convert $0.97\,cm^3$ into mm^3.

$$1\,cm^3 = 1000\,mm^3$$
$$\therefore 0.97\,cm^3 = 1000 \times 0.97\,mm^3$$
$$= 970\,mm^3$$

> Your answer will be a bigger number, so **multiply**.

Convert the following.

1 $3\,m^3 =$ _____

   $=$ _____ cm^3

2 $4.26\,cm^3 =$ _____

   $=$ _____ mm^3

3 $9\,800\,cm^3 =$ _____

   $=$ _____ m^3

4 $0.0051\,m^3 =$ _____

   $=$ _____ cm^3

5 $13\,572\,mm^3 =$ _____

   $=$ _____ cm^3

6 $2.37\,L =$ _____

   $=$ _____ cm^3

7 $19\,cm^3 =$ _____

   $=$ _____ L

8 $0.0078\,m^3 =$ _____

   $=$ _____ cm^3

9 A swimming pool is 9 m by 7 m and is 1.7 m deep. How many litres of water does it contain when full?

   _____

10 How many $250\,cm^3$ glasses will a 3 L bottle of milk fill?

   _____

Two-dimensional shapes: revision

- **Perimeter** is the distance around the outside of a two-dimensional (2D) shape.
- **Area** is the size of a flat surface inside a two-dimensional shape.

Complete the table to make your own list of formulae.

Shape	Perimeter	Area
Rectangle		Area = bh (base x height)
Triangle		
Circle		
Parallelogram and rhombus		
Trapezium		

Calculate the perimeters and areas of these shapes.

1

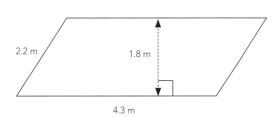

2.2 m   1.8 m   4.3 m

Perimeter = _____

= _____

Area = _____

= _____

= _____

2

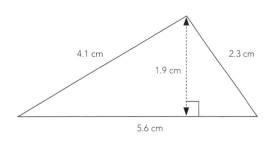

4.1 cm   1.9 cm   2.3 cm   5.6 cm

Perimeter = _____

= _____

Area = _____

= _____

= _____

3

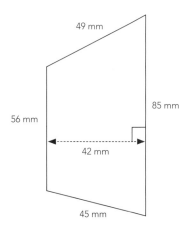

49 mm   85 mm   56 mm   42 mm   45 mm

Perimeter = _____

= _____

Area = _____

= _____

= _____

4

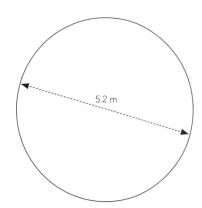

5.2 m

Perimeter = _____

= _____

Area = _____

= _____

= _____

Compound shapes and shapes with holes

- Compound shapes are made up of two or more other simple shapes.
- Carefully select the values that you use for your calculations. Some may not be needed, others will need to be calculated or their units changed to get the correct answer.

Examples:

1 This shape is made up of a rectangle and a semicircle.

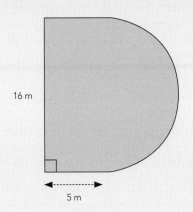

16 m

5 m

$$\text{Perimeter} = 16 + 5 + 5 + \frac{1}{2}(\pi \times 16)$$

$$= 51.13 \text{ m (2 dp)}$$

When combining areas, **add** the smaller areas together.

$$\text{Area} = 16 \times 5 + \frac{1}{2}(\pi \times 8^2)$$

$$= 180.53 \text{ m}^2 \text{ (2 dp)}$$

2 This shape is made up of a triangle with a square shape cut out of it.

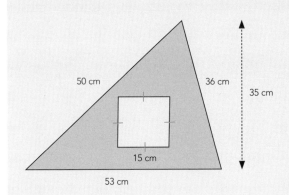

50 cm   36 cm   35 cm

15 cm

53 cm

$$\text{Total perimeter of figure} = (36 + 50 + 53) + (4 \times 15)$$

$$= 199 \text{ cm}$$

$$\text{Area} = \text{area triangle} - \text{area square}$$

$$= \frac{1}{2}(53 \times 35) - 15^2$$

Subtract the values when there is a hole.

$$= 702.5 \text{ cm}^2$$

Calculate the perimeters and/or areas of these shapes.

54

1

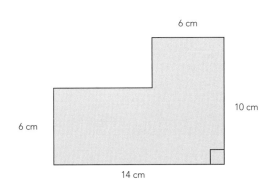

6 cm

10 cm

6 cm

14 cm

Perimeter = _____

= _____

Area = _____

= _____

= _____

2

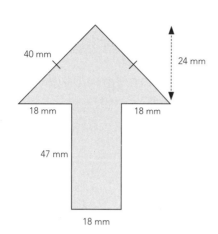

Perimeter = _____

= _____

Area = _____

= _____

= _____

3

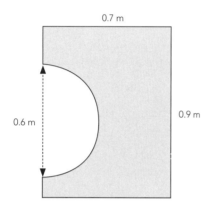

Perimeter = _____

= _____

Area = _____

= _____

= _____

4

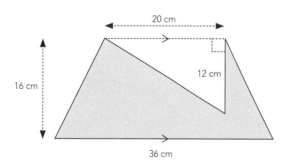

Area = _____

= _____

= _____

= _____

5

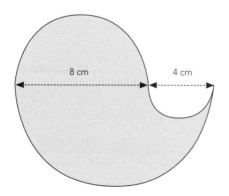

Perimeter = _____

= _____

Area = _____

= _____

= _____

= _____

 ISBN: 9780170477468

Three-dimensional shapes: surface area and volume

Complete the table to make your own list of formulae.

Shape	Volume	Surface area
Triangular prism		
Cylinder		
Cone		
Pyramid		
Sphere		

Surface area

- The surface area of an object is the sum of the areas of all its faces and surfaces.
- Drawing the net of the shape can be helpful.
- Don't forget to use square units in your answers.

Examples:

Cylinder

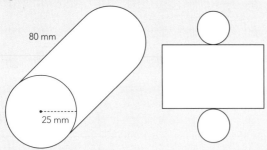

Surface area = area of curved side + 2(area of end)

$$= (80 \times \pi \times 50) + 2(\pi \times 25^2)$$

$$= 16\ 493\ \text{mm}^2 \ (0\ \text{dp})$$

When calculating surface area, the units must be squared.

Triangular prism

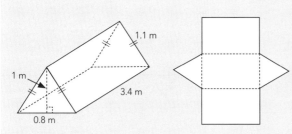

Surface area = 2(area side) + area base + 2(triangular end)

$$= 2(3.4 \times 1.1) + (0.8 \times 3.4) + 2(\frac{1}{2} \times 0.8 \times 1)$$

$$= 11\ \text{m}^2$$

Pyramid

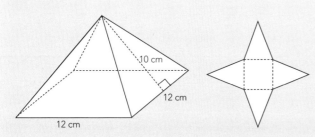

Surface area = area base + 4(area side)

$$= (12 \times 12) + 4(\frac{1}{2} \times 12 \times 10)$$

$$= 384\ \text{cm}^2$$

Sphere

Surface area $= 4\pi r^2$

$$= 4 \times \pi \times 1.4^2$$

$$= 24.63\ \text{cm}^2 \ (2\ \text{dp})$$

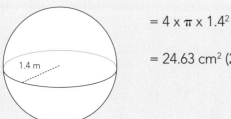

Cone

Surface area $= \pi r^2 + \pi r l$

$$= \pi \times 43^2 + \pi \times 43 \times 97$$

$$= 18\ 912\ \text{mm}^2 \ (0\ \text{dp})$$

Calculate the surface areas of the following shapes.

1

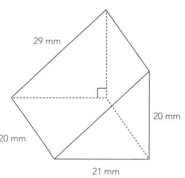

29 mm
20 mm
20 mm
21 mm

Net

_____

_____

_____

_____

2

27 cm
12 cm

Net

_____

_____

_____

_____

3

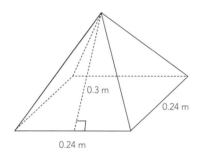

0.3 m
0.24 m
0.24 m

Net

_____

_____

_____

_____

4

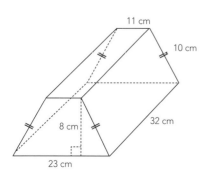

11 cm
10 cm
8 cm
32 cm
23 cm

Net

_____

_____

_____

_____

5

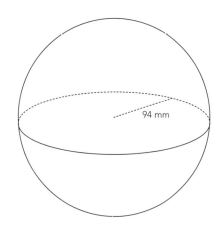

94 mm

_____

_____

_____

_____

6

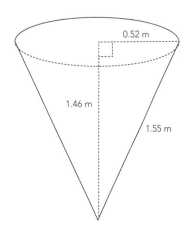

0.52 m

1.46 m

1.55 m

_____

_____

_____

_____

7

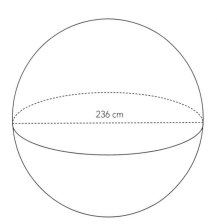

236 cm

_____

_____

_____

_____

8

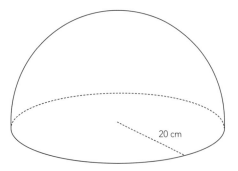

20 cm

_____

_____

_____

_____

Volume

- The volume of a three-dimensional (3D) shape is the amount of space the shape occupies.
- Don't forget to use cubic units in your answers.

Cuboids, prisms and cylinders

Cuboid

- A cuboid is a box shape.
- A cube is a box shape for which all the dimensions are equal.

$$\text{Volume} = \text{height} \times \text{width} \times \text{depth}$$
$$= h \times w \times d$$

Example:

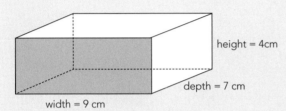

height = 4cm
depth = 7 cm
width = 9 cm

Volume = area of face x depth

$$= (4 \times 9) \times 7$$

$$= 252 \text{ cm}^3$$

> When calculating volume, the units must be cubed.

Prism

- A prism is a shape that has two identical parallel faces and flat sides.
- All of its cross-sections are identical.
- Cuboids are also prisms.

$$\text{Volume} = \text{area of face} \times \text{depth}$$

Example:

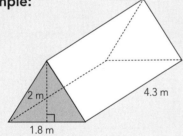

2 m
4.3 m
1.8 m

Volume = area of face x depth

$$= (\frac{1}{2} \times 1.8 \times 2) \times 4.3$$

$$= 7.74 \text{ m}^3$$

> Remember, the area of a triangle $= \frac{1}{2} \times$ base x height.

Cylinder

- Remember, a prism is a shape that has two identical parallel faces and flat sides.
- Cylinders are not prisms, but their volume can be calculated in the same way.

$$\text{Volume} = \text{area of face} \times \text{depth}$$
$$= \pi r^2 \times \text{depth}$$

Example:

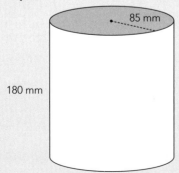

85 mm
180 mm

> **Depth** can be replaced by **length** or **height**, depending on the orientation of the cylinder.

Volume = area circle x height

$$= (\pi \times 85^2) \times 180$$

$$= 4\,085\,641 \text{ mm}^3 \text{ (0 dp)}$$

Calculate the volumes of these shapes. Round your answers to 4 sf.

1

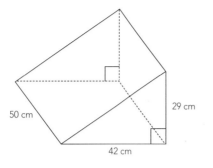

50 cm

29 cm

42 cm

_____

_____

2

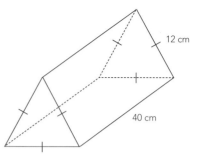

12 cm

40 cm

_____

_____

3

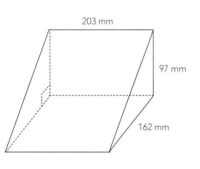

203 mm

97 mm

162 mm

_____

_____

4

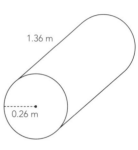

1.36 m

0.26 m

_____

_____

5

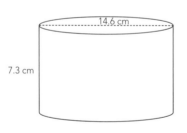

14.6 cm

7.3 cm

_____

_____

6

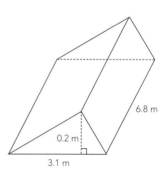

6.8 m

0.2 m

3.1 m

_____

_____

7

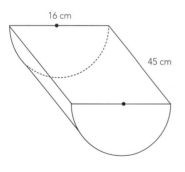

16 cm

45 cm

_____

_____

8

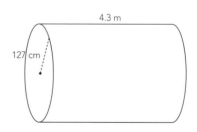

4.3 m

127 cm

_____

_____

Pyramids, cones and spheres

Pyramid
- A pyramid can have any polygon as its base, and its sides are triangles.

$$\text{Volume of pyramid} = \frac{1}{3} \times \text{(base area)} \times \text{height}$$

Example:

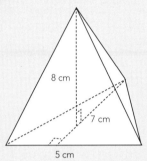

> You must use the **vertical** height.

Volume $= \dfrac{1}{3} \times$ (base area) $\times$ vertical height

$\qquad = \dfrac{1}{3} \times (\dfrac{1}{2} \times 7 \times 5) \times 8$

$\qquad = 46.\dot{6}$

$\qquad = 46.7$ cm³ (1 dp)

> Remember to round appropriately.

Cone
- A cone is similar to a pyramid, but it has a circular base and the sides are curved.

$$\text{Volume of cone} = \frac{1}{3} \times \text{(base area)} \times \text{height}$$

$$= \frac{1}{3} \pi r^2 h$$

> The base is always a circle.

Example:

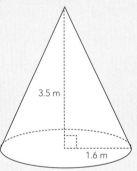

Volume $= \dfrac{1}{3} \times$ (base area) $\times$ vertical height

$\qquad = \dfrac{1}{3} \times (\pi \times 1.6^2) \times 3.5$

$\qquad = 9.38$ m³ (2 dp)

> Remember to use the radius, not the diameter.

Sphere
- A basketball is the shape of a sphere.

$$\text{Volume of sphere} = \frac{4}{3} \pi r^3$$

Example:

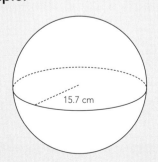

Volume $= \dfrac{4}{3} \times \pi \times 15.7^3$

$\qquad = 16\,210$ cm³ (0 dp)

Calculate the volumes of these shapes. Round your answers to 4 sf.

9

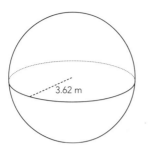

3.62 m

_____

_____

10

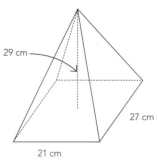

29 cm

27 cm

21 cm

_____

_____

11

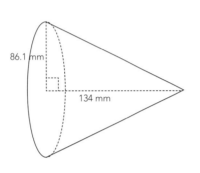

86.1 mm

134 mm

_____

_____

12

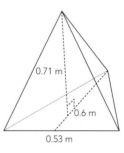

0.71 m

0.6 m

0.53 m

_____

_____

13

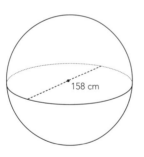

158 cm

_____

_____

14

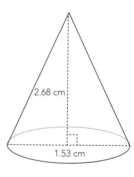

2.68 cm

1.53 cm

_____

_____

15

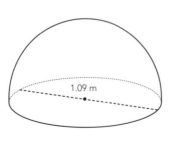

1.09 m

_____

_____

16

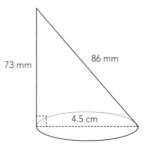

73 mm

86 mm

4.5 cm

_____

_____

Compound shapes: surface area and volume

- Compound shapes are shapes that are made up of other simple shapes.
- You will need to find the area or volume of the separate shapes and combine them.
- Sometimes, there are several ways of calculating these.
- If rounding is needed, do not do it until the end of your calculations.

Example:

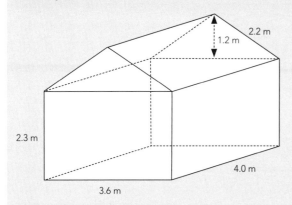

Area:

Area of floor	$= 4.0 \times 3.6$
	$= 14.4 \text{ m}^2$
Area of side walls	$= 2(2.3 \times 3.6 + 4.0 \times 2.3)$
	$= 34.96 \text{ m}^2$
Area of end triangles	$= 2(\frac{1}{2} \times 3.6 \times 1.2)$
	$= 4.32 \text{ m}^2$
Area of roof	$= 2(2.2 \times 4.0)$
	$= 17.6 \text{ m}^2$
Total area	$= 71.28 \text{ m}^2$

Volume:

Volume of triangular prism $= \frac{1}{2} \times 3.6 \times 1.2 \times 4.0$

$= 8.64 \text{ m}^3$

Volume of cuboid $= 3.6 \times 2.3 \times 4.0$

$= 33.12 \text{ m}^3$

Total volume $= 8.64 + 33.12$

$= 41.76 \text{ m}^3$

Calculate the total surface area and volume of each shape. Round your answers to 4 sf.

1   This shape is made up of a square-based cuboid with a square-based pyramid.

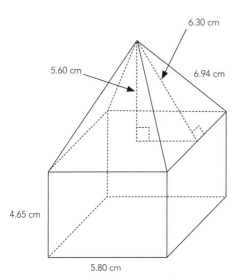

_____

_____

_____

_____

_____

_____

_____

_____

_____

_____

2 This shape is a cuboid with a cylindrical hole.

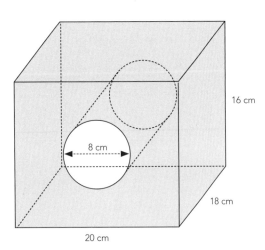

16 cm

8 cm

18 cm

20 cm

_____

_____

_____

_____

_____

_____

_____

_____

_____

_____

_____

_____

3 This shape is made up of a hemisphere and a cone.

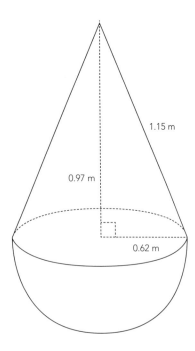

1.15 m

0.97 m

0.62 m

_____

_____

_____

_____

_____

_____

_____

_____

_____

_____

_____

4   This shape is made up of a cylinder and a cone.

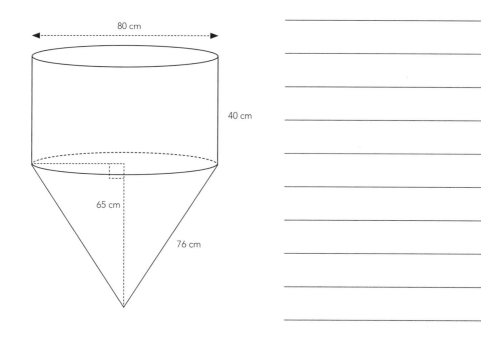

_____

_____

_____

_____

_____

_____

_____

_____

_____

_____

_____

_____

_____

5   This shape is a cube with a pyramid-shaped hole.

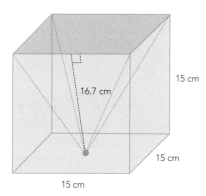

_____

_____

_____

_____

_____

_____

_____

_____

_____

_____

_____

_____

_____

Working backwards

- Sometimes, you can use the formula for an area or volume to calculate an unknown dimension.

Examples:

1 The curved surface of a cylinder is a 50 cm square when it is unrolled. Calculate its radius.

$$\text{Circumference of the cylinder} = 2 \times \pi \times r = 50 \text{ cm}$$

$$r = \frac{50}{2 \times \pi}$$

$$r = 7.96 \text{ cm (2 dp)}$$

2 The volume of a sphere is 904.8 cm³. Calculate its radius.

$$\text{Volume of sphere} = \frac{4}{3}\pi r^3 = 904.8 \text{ cm}^3$$

$$r^3 = \frac{904.8 \times 3}{4\pi}$$

$$r = \sqrt[3]{\frac{904.8 \times 3}{4\pi}}$$

$$r = 6 \text{ cm (0 dp)}$$

3 The surface area of this open box (no lid) is 444 cm³.

Calculate length d.

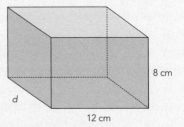

$$\text{Surface area} = \text{base} + 2(\text{front wall}) + 2(\text{side wall}) = 444 \text{ cm}^3$$

$$12d + 2(12 \times 8) + 2(8d) = 444$$

$$28d + 192 = 444$$

$$28d = 252$$

$$d = 9 \text{ cm}$$

You could check your answer by working forwards.

Answer the following questions.

1 The internal dimensions of the base of this box are 15 cm by 20 cm. The box holds 2.4 L. Calculate its internal depth.

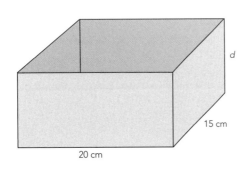

2 The volume of this figure is 8960 cm³. Calculate its width.

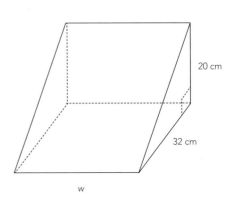

_____

_____

_____

_____

3   The volume of this pyramid is 270 cm³. Calculate its vertical height.

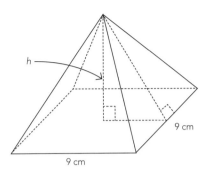

9 cm

9 cm

_____

_____

_____

4   The area of a circle is 0.7854 m². Calculate its radius.

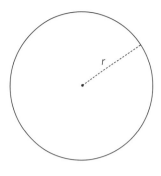

_____

_____

_____

5   This cone has a radius of 20 cm and a volume of 10 472 cm³. Calculate its height.

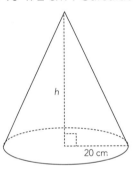

20 cm

_____

_____

_____

6   This hemisphere has a total surface area of 15 843 cm². Calculate its diameter.

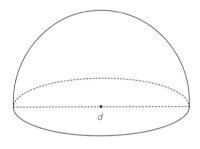

d

_____

_____

_____

7   This block of wood is 78 mm high and 62 mm deep. Its total surface area (including the base) is 24 792 mm². Calculate its width.

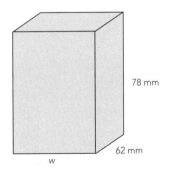

78 mm

62 mm

w

_____

_____

_____

8   This shape is a cube with a square-based pyramid on top. The cube has sides of 1.2 m. The total surface area of the shape is 8.16 m². Calculate the height of each face of the pyramid.

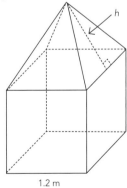

1.2 m

_____

_____

_____

Estimation

There are four stages to estimation:

1 Decide what **model** best fits the shape, and write down the relevant **formula**.

2 Estimate values for the **variables** in the formula and justify your estimate.

3 **Calculate** the estimation required and round it to a maximum of two significant figures.

4 Discuss **assumptions** and limitations of your model, and their impacts on your answer.

Example: Estimate the amount of ice cream needed to fill the cone and add a scoop on top.

8 cm

3.5 cm

1 To fill the cone

Model: cone.   Formula: $V = \frac{1}{3}\pi r^2 h$

Variable(s):   height (h): 7 cm because the point of the cone will be mostly wafer.

radius (r): 1.5 cm because we need to account for the cone thickness.

Estimate of volume: $V = \frac{1}{3}\pi (1.5)^2 \times 7$

$= 16 \text{ cm}^3$ (0 dp)

> Because you are estimating, **round** the answer **appropriately**. 17 cm³ would also have been an acceptable estimation.

2 The scoop

Model: hemisphere.   Formula: $V = \frac{1}{2}\left(\frac{4}{3}\pi r^3\right) = \frac{2}{3}\pi r^3$

Variable(s):   radius (r): 2 cm because the scoop has a slightly bigger radius than the cone.

Estimate of volume: $V = \frac{2}{3}\pi \times (2)^3$

$= 17 \text{ cm}^3$ (0 dp)

> Justify your choice of values for the variables.

Estimate of total ice cream needed: 16 + 17 = 33 cm³.

> You will **always** have to assume that the object is an exact geometric shape, but it almost certainly won't be.

Assumptions:

1 The cone is an exact cone shape with, for instance, straight sides. The thickness of wafer forming the cone is consistent, but because it is wrapped, it won't be.

2 The scoop is a perfect hemisphere, which it almost certainly will not be. The surface of the ice cream will be lumpy, and if the scoop is heaped up before it is added to the cone, it will probably be bigger than a hemisphere. I have also assumed that there are no air gaps in the ice cream.

Limitations:

The thickness of the cone limits the amount of ice cream, but there is no limit to how tall the scoop can be.

       ISBN: 9780170477468

Answer the following questions.

1 Estimate the volume of tomato sauce needed to fill this container. Its total width is 9.5 cm.

Model: _____

Formula: _____

Variable: r = _____ because _____

Estimate of volume = _____

= _____

Assumption(s): _____

_____

_____

Limitation(s): _____

_____

2 Estimate the volume of soil in this pile. The shovel beside it is 1.5 m long.

Model: _____

Formula: _____

Variables: _____ = _____ because _____

_____ = _____ because _____

Estimate of volume = _____

= _____

Assumption(s): _____

_____

_____

Limitation(s): _____

_____

3 Estimate the volume of milk in this 25 cm high jug.

25 cm

Model: _____

Formula: _____

Variables: _____ = _____ because _____

_____ = _____ because _____

Estimate of volume = _____

= _____

Assumption(s): _____

_____

_____

Limitation(s): _____

_____

4 This puddle of paint has a 16 cm long brush beside it. Estimate the volume of the spilt paint.

_____

_____

_____

_____

_____

_____

_____

_____

  ISBN: 9780170477468

5 Estimate the volume of drink in this glass. Its rim is 8 cm across.

8 cm

_____

_____

_____

_____

_____

_____

_____

_____

_____

_____

6 The part of a standard one-litre milk carton that can contain milk has a base which is 7.3 cm square, and it is 18 cm tall.

18 cm

Calculate the volume of the cuboidal part of the carton. _____

_____

Suggest a reason why your calculated volume is less than a litre.

_____

_____

_____

_____

Applications

Example:
Terri has a rectangular concrete slab in his woolshed. He is going to put a water cylinder on the slab. The slab measures 110 cm by 86 cm.

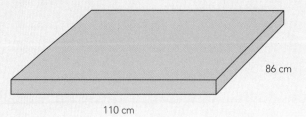

The water cylinder must:
- hold a minimum of 1000 L
- have the maximum radius possible without overhanging the slab
- have the minimum height possible, but this must be a whole number of centimetres.

Calculate the height of the cylinder.

$1000 \text{ L} = 1\,000\,000 \text{ cm}^3$

Maximum radius that will fit on the slab = 43 cm

Volume of the cylinder = $\pi \times r^2 \times h = 1\,000\,000 \text{ cm}^3$

$$h = \frac{1\,000\,000}{\pi \times 43^2}$$

$h = 172.15 \text{ cm (2 dp)}$

So the water cylinder must be 173 cm high.

Answer the following questions.

1 The school caretaker is painting courts for the game Four Square. The courts are 2.35 metres square, and the painted lines are 5 cm wide. In order to work out how much paint he must buy, he needs to know how much area is taken up by the lines. Calculate the area of the painted lines.

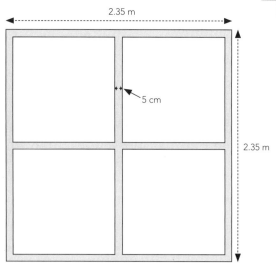

_____

_____

_____

_____

2 Hamish is making a concrete water trough for his animals. It will be 1 m wide, 60 cm deep and 45 cm high. The 'walls' will be 5 cm thick.

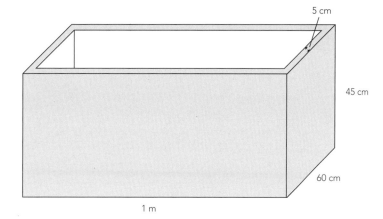

5 cm

45 cm

60 cm

1 m

a Calculate the volume of concrete needed to make the floor and walls of the trough.

_____

_____

_____

b Bags of concrete mix have a mass of 20 kg and each bag makes 10 L of concrete. How many bags of concrete will he need?

_____

_____

c He is going to put two coats of paint on the outsides of the walls and the top lip of the trough (the green area). Calculate the area (in square metres) of the painted surface.

_____

_____

_____

d One litre of paint covers 11 square metres. Calculate the volume of paint needed for the two coats.

_____

_____

e He fills the trough to within 5 cm of the top. How many litres of water will it contain?

_____

_____

_____

3   The side and the circular ends of a cylinder are to be cut from one piece of a steel that is 100 cm wide, as shown. Calculate your answers in centimetres to 2 dp.

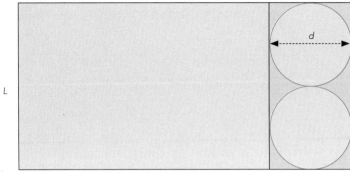

100 cm

a   Calculate the maximum diameter for the cylinder.

_____

_____

b   Calculate *L*, the length of steel needed to make the cylinder.

_____

c   Calculate the area of steel (the grey area) that will be left over.

_____

_____

4   Consider the ice cream on page 138. Assume that the diameter of the scoop is 4 cm.

a   The ice cream will be dipped in melted chocolate to produce a layer 2 mm thick on the outside of the scoop.

Calculate the volume of chocolate needed.

_____

_____

_____

b   Assume the outer diameter of the cone is 3.5 cm and its vertical height is 8 cm. The cone is, on average, 1.5 mm thick. Calculate the volume of the wafer making up the cone.

_____

_____

_____

 Challenges

1 Mariah is making a square-based pyramid paperweight out of glass. A cubic metre of glass has a mass of 2500 kg.

What will the mass of her paperweight be?

8 cm

7 cm

_____

_____

_____

_____

_____

_____

_____

_____

_____

_____

Assumptions: Describe any assumptions you have made in the calculations of your answer. Describe the impact these might have on your model and your answer.

_____

_____

_____

_____

Limitations: Describe any limitations of your answer.

_____

_____

Generalisation: Write a formula which could be used to find the mass of any square-based pyramidal glass paperweight, given the dimensions of the base (*b*) and its vertical height (*h*).

_____

_____

_____

2   Manu is training for a 400 m race.
      • He runs a warm-up lap of a 400 m track at x metres per second.
      • Then he does a second lap at four times the speed of his first lap.
      • He takes a total of 320 s to run the two laps.

      How long does he take to run the second lap and how fast was he running?

_____

_____

_____

_____

_____

_____

_____

_____

_____

Assumptions: Describe any assumptions you have made in the calculations of your answer.
Describe the impact these might have on your model and your answer.

_____

_____

_____

_____

Limitations: Describe any limitations of your answer.

_____

_____

_____

3 Alex is making a water trough which is shaped like a triangular prism. The ends will be equilateral triangles. The triangle sides on the **inside** of the trough will be 70 cm long. He needs the trough to hold 500 litres of water. Calculate the internal length (*L*) of the trough.

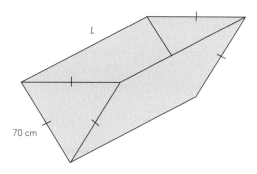

_____

_____

_____

_____

_____

_____

Assumptions: Describe any assumptions you have made in the calculations of your answer. Describe the impact these might have on your model and your answer.

_____

_____

_____

Limitations: Describe any limitations of your answer.

_____

_____

_____

Generalisation: Write a formula for the volume of any trough which is the shape of a prism with equilateral triangles for its ends. Let the *t* be the side of the triangles and *L* be the length of the trough.

_____

_____

_____

4   Earth's distance from the moon varies, but the average distance is 3.825×10^5 km.
    If a rocket travels to the moon at an average speed of 5.0×10^4 km/hour, how long will it take
    to get there?

_____

_____

_____

_____

_____

_____

Assumptions: Describe any assumptions you have made in the calculations of your answer.
Describe the impact these might have on your model and your answer.

_____

_____

_____

Limitations: Describe any limitations of your answer.

_____

_____

_____

Generalisation:

_____

_____

_____

   ISBN: 9780170477468

5 A wobbly toy consists of a hemisphere with a cone on top. The cone has been slightly rounded in case a child falls on top of the toy.

- The diameter of both the cone and the sphere is 16 cm.
- The overall height of the toy is 20 cm.
- The entire surface of the toy needs two coats of paint.
- One litre of paint covers 6 m^2.
- The paint comes only in 2 L tins.

How many of these toys could receive two coats of paint from a 2 L tin?

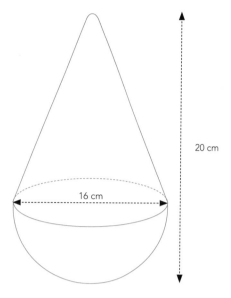

20 cm

16 cm

Practice tasks

Practice task one

When solving this problem you need to:
- Show calculations, using correct mathematical statements.
- Explain what you are calculating at each stage of your solution.
- Provide reasoning which is linked to the context.

Information

Paper wasps arrived in New Zealand in the late 1970s. They have become a problem because they compete with our native insects, birds, bats and lizards for live insect prey.

Beatrice has found many paper wasp nests on her property:
- The area of Beatrice's property is 8500 m². Percy estimates that the wasp nests on Beatrice's property are at a density (*D*) of 300 nests per hectare.

- Wasp nests are approximately spherical. The average diameter of the nests on her property is 4.5 cm.

- The volume (*V*) of spray needed (*h*) is given by this formula:
$$V = \frac{ADS \times 10^{-2}}{3}$$
where A stands for area (Ha), D stands for density (nests per Ha) and S stands for average surface area of nests (cm²).

- In order to reach the biggest nest, Percy needs to place his ladder against a wall. Site safe recommends an angle of 75° between the ground and the ladder. In order to access the nest, he needs the ladder to reach at least 3.5 m up the wall.

- Percy estimates that it will take about 5 hours to spray the nests on Beatrice's property.

- This is an advertisement for Percy's Pest Services. The charges of a competitor, Ernie's Exterminators, are on the graph.

- Beatrice suggests Percy charges a set fee of $25 plus $65 an hour.

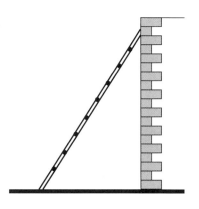

Percy's Pest Services

$368 including GST for complete extermination.

Prompt payment discount: we pay the GST.

Task instructions
- Estimate the number of nests on Beatrice's property.
- Calculate the minimum amount of spray needed to spray all wasp nests on Beatrice's property.
- Calculate the minimum length of ladder that Percy would need to use.
- Compare the amount that Ernie would charge for 5 hours' work, with Percy's advertised price for the same job.
- Percy changes his method for calculating the amount he charges to Beatrice's suggestion. Add this to the graph and describe the circumstances under which he would charge less than Ernie.
- Discuss any assumptions you have made, any limitations and the impacts they have on your solutions.

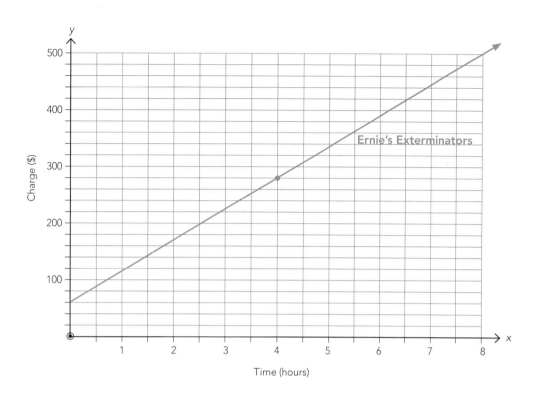

_____
_____
_____
_____
_____
_____
_____
_____
_____
_____
_____
_____
_____
_____
_____
_____
_____
_____
_____
_____

ISBN: 9780170477468

Practice task two

When solving this problem you need to:
- Show calculations, using correct mathematical statements.
- Explain what you are calculating at each stage of your solution.
- Provide reasoning which is linked to the context.

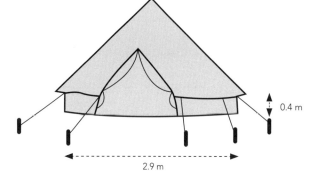

Information
- Hemi's canvas tent has a cylindrical base with a conical top.

- Internally, the tent has a diameter of 2.9 m.

- It is recommended that for sleeping in a tent, each person needs a minimum area of 1.4 m².

- The conical section of the tent needs to be waterproofed. Its diameter is 3.2 m, and its vertical height is 1.9 m. It takes 50 mL of spray to cover each square metre of canvas.

- In order to make up the waterproof spray, the waterproofing concentrate needs to be mixed with water in a ratio of 3:2.

- In order to stabilise the tent, guy ropes are attached from the top of the cylindrical base to pegs around the tent. The guy ropes attach to the pegs at points 10 cm above ground level. It is recommended the guy ropes have angles of elevation of less than 13°.

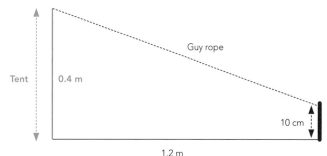

- Hemi needs to purchase inflatable mattresses. They are usually $120, but Better Bargains will reduce the price by 15%. At Winnie's Warehouse the full price is also $120, but they say that they will pay the GST.

Task instructions
- Calculate the recommended maximum number of people who should sleep in this tent.
- Calculate how much spray Hemi will need to waterproof the tent.
- Describe how he should make up the waterproof spray.
- Will Hemi's placement of the pegs for guy ropes mean he complies with the recommended angle of elevation? What adjustments would he need to make in order to comply?
- Which outlet gives Hemi the best deal for his inflatable mattresses?
- Discuss any assumptions you have made, any limitations and the impact they have on your solutions.

ISBN: 9780170477468

Practice task three

When solving this problem you need to:
- Show calculations, using correct mathematical statements.
- Explain what you are calculating at each stage of your solution.
- Provide reasoning which is linked to the context.

Information

Kiri is buying a tiny house. Its floor plan and orientation are shown.
- The bed is exactly 2 m wide in total.

- During construction, a brace is needed to support the framing for the bathroom.

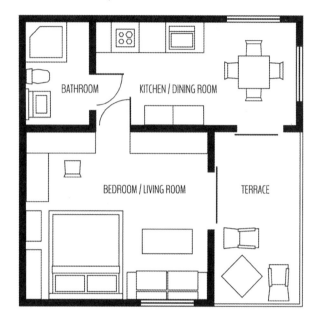

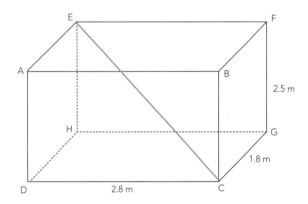

- The roof will be made from sheets of corrugated iron which are 4.0×10^{-4} m thick. When the delivery truck arrives, it has a stack of sheets of corrugated iron which is 20 cm high.

- The terrace has an area of 12 m². Every litre of paint covers 10 square metres. He needs to apply two coats. The paint comes in three sizes:
    - 1 L costs $59.99
    - 2 L costs $89.99
    - 4 L costs $142.00

- Kiri is building a rectangular vegetable garden along the side of the house which borders the bedroom and the bathroom. She has 12 m of timber edging, and would like the garden area to be as large as possible.

Task instructions
- Calculate the internal length of the kitchen/dining room.
- How much timber will be needed for the brace?
- Make a recommendation for how Kiri should buy the paint for her terrace floor.
- How many sheets of corrugated iron were on the truck?
- Help Kiri to optimise the size of her vegetable garden.
- Discuss any assumptions you have made, any limitations and the impact they have on your solutions.

 ISBN: 9780170477468

Width	Length	Area

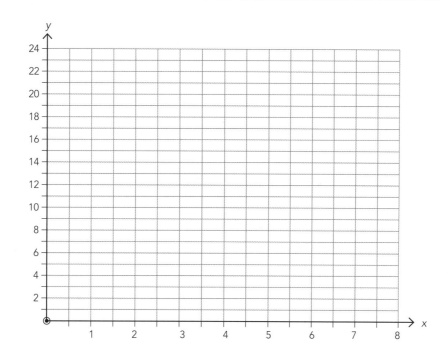

_____

_____

_____

_____

_____

_____

_____

_____

_____

_____

ISBN: 9780170477468

Practice task four

When solving this problem you need to:
- Show calculations, using correct mathematical statements.
- Explain what you are calculating at each stage of your solution.
- Provide reasoning which is linked to the context.

Information

Marama is in charge of making poi for the kapa haka group. She needs to make 80 pairs of poi.

Each poi is approximately spherical with a diameter of 8 cm. It is made from a circle of fabric stretched around some stuffing.

8 cm

Each cord is 21 cm long and made up of 12 strands of carpet yarn (wool) which are braided together.

- The cords for the poi will be braided using 12 strands of carpet yarn (wool). The ratio of the length of each strand of carpet yarn to the length of finished cord (including the tassel) is 5:3. Each finished cord will be 21 cm long.

- A cone of carpet yarn contains 600 m of yarn. A cone costs $41.40 including GST, but because it's for kapa haka, the retailer will pay the GST.

- The circular covering for each poi will need to be cut from a square of fabric. The circle needs to be large enough to cover its surface area, plus a 2 cm margin to allow for tying and connecting to the cord. The diameter of each poi will be 8 cm.

- The stuffing for each poi will come from old cushions. A cushion contains about 0.018 m³ of stuffing.

- Tickets to the local kapa haka competition cost $30 for adults and $18 for students and children. The total value of the 2263 tickets sold was $58 854.

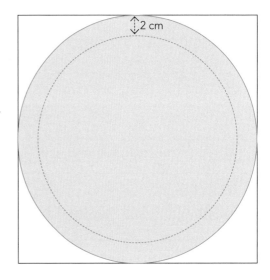

2 cm

Task instructions

- Calculate the cost of the carpet yarn needed to make the 80 pairs of poi.
- What area fabric will Marama need to make the 160 poi?
- How many cushions will be needed to stuff all the poi?
- How many adults and students attended the kapa haka competition?
- Discuss any assumptions you have made, any limitations and the impact they have on your solutions.

ISBN: 9780170477468

 # Answers

Number (pp. 6–33)

Rates and proportions (pp. 7–10)

1	$17.97	**2**	$83.65
3	18 jars	**4**	8 750 000 snaps
5	128 mm	**6**	$NZ900.00

Inverse proportions (p. 8)

7	5.5 minutes	**8**	8 m per hour
9	48 min	**10**	9 days
11	4 hours 54 minutes	**12**	14 min

Assumptions and limitations (pp. 9–10)

1 a 35 cookies

b We assume that the cookies will be the same size and shape. This result would apply only if the same recipe is used.

2 a 33 000 cm or 330 m

b We have assumed that she and the flea have similar bodies, and that their mass to height ratios are similar. Fleas have their skeletons on the outside but people have theirs on the inside, which changes the biomechanics involved.

3 a 40 minutes

b We have assumed that the 1000-word document is similar to the document used to calculate his rate of typing. The answer would not apply to scientific documents with numbers and technical terms, documents with complex layouts or documents in a foreign language.

4 a $961.54

b We must assume that the exchange rate is the same when he arrives in England, and that the exchange bureau doesn't charge a fee for exchanging money.

5 a 88 200 spiders

b We have to assume that spider densities are similar on playing fields as in the green areas on which the average was based. In green areas with bushes and trees, there would be lots of places for spiders to live, but playing fields have no tall plants and are constantly trampled, so it is likely that there will be fewer spiders than calculated.

6 a 2 days and 10 hours

b We must assume that conditions for circling the globe are similar to those when flying the length of New Zealand. This is unlikely because birds circling the globe probably take advantage of air currents which are in the same direction. New Zealand is exposed to westerly winds, which would be less help when flying its length. It is probably only certain species of albatross that this would apply to.

Ratios (pp. 11–14)

Using ratios where the total is given (pp. 11–12)

1	10 kg:6 kg	**2**	$66:$24
3	$48:$60	**4**	144 (not 143.7)
5	59 400	**6**	3.328 million
7	$97.50:$130.00:$32.50		
8	4.8 kg:3.2 kg:3.2 kg:1.6 kg		
9	Mila 15, Matiu18		
10	3 kg:4 kg:2 kg:1 kg		

Ratio calculations where one part is given (pp. 13–14)

1	48 m	**2**	5.3 mm
3	45 000 species	**4**	7.8 million
5	166.4 cm	**6**	258 g flour, 86 g water
7	$270	**8**	62.98 m

Scale diagrams (pp. 15–19)

1	1:20	**2**	1:400 000
3	40:1	**4**	20:1

You may get different answers to these questions. If so, check with your teacher.

5	2 m	**6**	750 cm or 7.5 m
7	47.5 cm	**8**	12 mm or 1.2 cm
9	3.3 mm	**10**	12.6 km
11	33.5 km		

12 a 5.5 m

b No. The bedroom is 3.0 cm on the diagram, which is only 1.50 m.

13 a 1:100   **b** 9.1 m

Powers (pp. 20–21)

Fractional powers and roots (p. 20)

1	4	**2**	7
3	0.2	**4**	5
5	0.3	**6**	12
7	2	**8**	0.4

Negative powers (p. 21)

1	$\frac{1}{2}$ or 0.5	**2**	$\frac{1}{100}$ or 0.01
3	125	**4**	$\frac{16}{9}$

Combining negative and fractional powers (p. 21)

1	$\frac{1}{4}$	**2**	$\frac{1}{3}$
3	$\frac{1}{100}$	**4**	11
5	$\frac{3}{2}$ or 1.5	**6**	$\frac{5}{2}$ or 2.5
7	4	**8**	1000
9	5.28	**10**	1.79

Percentages (pp. 22–26)

Using percentages (pp. 22–23)

1	22.75 km	2	$109.20
3	712.5 mL	4	$198.10
5	62 km	6	$67.50
7	86.25 kg	8	691.2 g
9	$360.00	10	245 mL
11	21 games	12	118 students
13	$25.20	14	$13 950.00
15	$7475.00	16	$349.93
17	15.1% (1 dp)	18	0.0001%

Calculating percentage changes (p. 24)

1	+60%	2	–40%
3	–25%	4	+15%
5	+18%	6	–61% (1 dp)
7	25%	8	51.5% (1 dp)
9	43.4% (1 dp)	10	5.77% (2 dp)

GST (pp. 25–26)

1	$66.70	2	$900.45
3	$11 509.20	4	$593 400
5	$17.04	6	$13 956.52
7	$364.35	8	$72 173.91
9	$36.00	10	$1.14
11	$151.20	12	$6420.00
13	$586.95	14	76c
15	$138.26	16	$517.50
17	$33.78	18	$22.61
19	$151.80	20	$586.04
21 a	$280	b	$322

22 GST is only about 13% of the GST inclusive price. Example: Pre-GST price = $100. So the GST inclusive price = $115, and the GST = $15. The saving is $15 out of the GST *inclusive* price of $115.

So the percentage saving is $\frac{15}{115} \times 100 = 13\%$ (1 dp), not 15%.

Standard form (pp. 27–29)

Standard form to ordinary numbers (p. 27)

1	15 600	2	0.00391
3	521 000	4	0.094
5	2 700 000	6	0.00000823

Ordinary numbers to standard form (pp. 28–29)

1	4.1896×10^4	2	6.1×10^1
3	9.564×10^5	4	1.90×10^2
5	8×10^0	6	1.0×10^6
7	2.825×10^1	8	4.003×10^2
9	5×10^{-1}	10	7.1×10^{-4}
11	6×10^{-10}	12	1.011×10^{-4}
13	2.6×10^{-3}	14	1.01×10^{-4}
15	3 080 000 users	16	0.08 mm
17	0.00000055 m	18	1.28×10^7
19	1.12×10^{19}	20	1.0×10^9 or 1 billion
21	1.90×10^{27} kg	22	4.18×10^{13}
23	5.12×10^{11}	24	5.0×10^{-4} or 0.0005

Interest (pp. 30–33)

Simple interest (p. 30)

1

Value of loan	Rate of interest	Interest per year	Number of years	Total debt if none paid back
$800	4%	$32.00	5	$960.00
$21 000	3.5%	$735.00	6	$25 410.00
$352 000	7.2%	$25 344.00	3	$428 032

2	$900.00	3	$85 200.00
4 a	$330.00	b	$6490.00
5 a	$135.00	b	11 years

Compound interest (pp. 31–33)

1

Value of loan	Rate of interest	Number of years	Calculation	Total debt
$200	4%	5	200×1.04^5	$243.33
$18 000	5.5%	2	$18\,000 \times 1.055^2$	$20 034.45
$149 500	1.9%	6	$149\,500 \times 1.019^6$	$167 373.35

2

Value of loan	Rate of interest	Number of years	Compound period	Calculation	Total debt
$200	4%	5	Monthly	$200\left(1 + \frac{0.04}{12}\right)^{12 \times 5}$	$244.20
$18 000	5.5%	2	Fortnightly	$18\,000\left(1 + \frac{0.055}{26}\right)^{26 \times 2}$	$20 090.67
$149 500	1.9%	6	Weekly	$149\,500\left(1 + \frac{0.019}{52}\right)^{52 \times 6}$	$167 548.95

3	$8932.62	4	$143 024.61
5	0.89%	6	$5900.00

7

Money withdrawn at the end of	Scheme one	Scheme two
one year	$3135.00	$3090.00
two years	$3270.00	$3182.70
three years	$3405.00	$3278.18
four years	$3540.00	$3376.53

b 4.3%

c I assume that Kora did not withdraw any money during the four years. If she did, that would reduce the value of her investment.

8 a Scheme one: $16 390.91
Scheme two: $16 401.65

b 3.1%

c I assume that Sione did not withdraw any money during the three years. If he did, that would reduce the value of his investment.

Algebra (pp. 34–86)

Substitution (p. 35)

1	58	2	3
3	20	4	4
5	–16	6	–2
7	61	8	80π

9 $\dfrac{256}{3}\pi$ or $85\dfrac{1}{3}\pi$ **10** $64 - \dfrac{64}{3}\pi$

Linear equations (pp. 36–53)

Solving linear equations (pp. 36–37)

1 $x = 15$

2 $x = \dfrac{45}{7}$

3 $x = \dfrac{11}{4}$ or 2.75

4 $x = 1$

5 $x = -3$

6 $x = 4$

7 $x = 36$

8 $x = \dfrac{25}{8}$ or 3.125

9 $x = \dfrac{50}{8}$ or 6.25

10 $x = \dfrac{11}{8}$ or 1.375

11 $x = \dfrac{35}{2}$ or 17.5

12 $x = -4$

13 $x = \dfrac{11}{4}$ or 2.75

14 $x = \dfrac{-1}{3}$

15 $x = -8$

16 $x = 12$

Forming and solving linear equations (pp. 38–39)

1 $(s + 7) + (s + 7) + s = 29$
     $s = 5$
The shortest side is 5 cm long, the longer sides are 12 cm long.

2 $5(y + 2) + 3y = 22$
          $y = 1.50$
Small container costs $1.50, large container costs $3.50.

3 $2x + 2(x - 3) = 42$
          $x = 12$
The long sides are 12 cm, the short sides are 9 cm.

4 $45a + 18(12 - a) = 297$
          $a = 3$
Three adults and nine children went to the circus.

5 $2(7 + y) = 35 + y$
          $y = 21$
Kurt's dad will be double Kurt's age in 21 years' time.

Plotting linear equations (pp. 40–41)

1

x	y
0	15
1	13
2	11
3	9
4	7
5	5
6	3

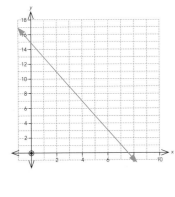

2

x	y
0	7
2	6
4	5
6	4
8	3
10	2

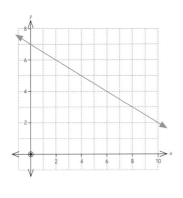

3

x	y
0	8
1	10
2	12
3	14
4	16
5	18

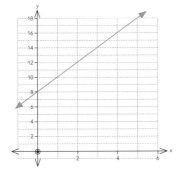

4

x	y
0	16
1	10
2	4
3	-2

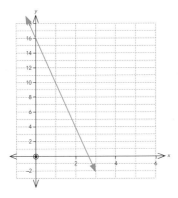

The gradient of a line (pp. 42–43)

1 **a** $m = 2$ **b** $m = 0$

    c $m =$ undefined   **d** $m = -3$

    e $m = -\dfrac{1}{6}$ **f** $m = \dfrac{4}{3}$

    g $m = \dfrac{2}{5}$ **h** $m = -\dfrac{2}{7}$

    i $m = -\dfrac{3}{2}$ **j** $m = \dfrac{1}{4}$

Using features of straight lines to write equations (pp. 44–45)

1 y intercept = 10
    Gradient = 3
    Equation $y = 3x + 10$

2 y intercept = 50
    Gradient = –4
    Equation $y = -4x + 50$

3 y intercept = 200
    Gradient = 25
    Equation $y = 25x + 200$

4 y intercept = 18

Gradient = $\dfrac{-1}{5}$

Equation $y = -\dfrac{1}{5}x + 18$

5 y intercept = 5

Gradient = 15

Equation $y = 15x + 5$

6 y intercept = 0

Gradient = $\dfrac{3}{10}$

Equation $y = \dfrac{3}{10}x$

7 y intercept = 19

Gradient = $\dfrac{-1}{3}$

Equation $y = \dfrac{-1}{3}x + 19$

8 y intercept = 85

Gradient = $-\dfrac{3}{4}$

Equation $y = -\dfrac{3}{4}x + 85$

Equations of vertical, horizontal and parallel lines (p. 46)

1 $y = 2$ **2** $x = -2$

3 $y = \dfrac{3}{2}x + 3$ **4** $y = -\dfrac{1}{2}x + 4$

Writing equations given two points (pp. 47–48)

1 $y = 4x + 3$ **2** $y = x + 11$

3 $y = -2x + 21$

4 $y = 0.5x - 5$ or $y = \dfrac{1}{2}x - 5$

5 $y = 0.75x + 32$ or $y = \dfrac{3}{4}x + 32$

6 $y = -0.5x + 43$ or $y = -\dfrac{1}{2}x + 43$

7 $y = 3.5x$ or $y = \dfrac{7}{2}x$

8 $y = -2.5x + 73$ or $y = -\dfrac{5}{2}x + 73$

9 a $y = 65x + 45$ **b** $45.00
    c $65.00

Applications (pp. 49–53)

1 a $80   **b** The y intercept is 80.
    c $30.00   **d** The gradient is 30.
    e $S = 30n + 80$ **f** 16 weeks

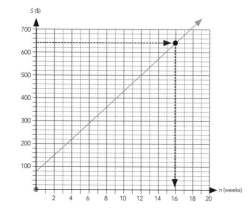

2 a $1   **b** The y intercept is 1.
    c $0.30   **d** The gradient is 0.30.
    e $C = 0.3t + 1$ **f** $7
    g 24 min
    h The y intercept would be at 2 rather than 1.
    l The gradient in both would be 0.30.
    j $C = 0.3t + 2$
    k $C = 0.2t + 4$

3 a $2.50   **b** The gradient is 2.5.
    c $0.75   **d** $45.00
    e $P = 2.5b - 30$ **f** 12
    g He would have lost $10.
    h Equation: $P = 2b - 30$
       He would need to sell 15 bags to break even.
       Selling all 40 bags would get him $50 profit.

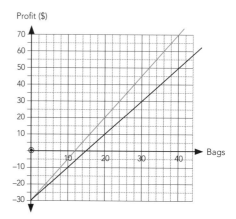

4 **a** and **c**

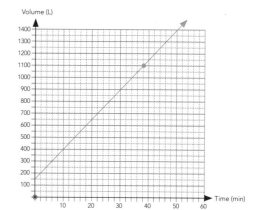

    b 1100 L   **d** 48 min
    e $V = 25t + 150$ **f** 56 min
    g $V = 25t + 300$
    h The intercept is higher at 300 L rather than 150 L. The gradient is 25 in both relationships.
    i $V = 15t + 200$

5 a 4.2 km
    b The y intercept.
    c $\dfrac{4200}{70} = 60$ m/min
         $= 0.06$ km/min
         $= 3.6$ km/hour
    d The gradient = –60.
    e 900 m
    f 50 min
    g $D = -60t + 4200$

ISBN: 9780170477468

h Gradient = –80

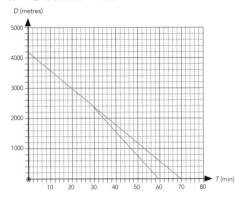

i She will arrive at the lake after 60 min rather than 70 min.

Piecewise functions (pp. 54–57)

1 a Section A shows a set charge of $35 for 100 minutes or fewer.

b $0.45 per minute

2 a $10.00 **b** $20.00

3 a $D = 18T$

$D = 18$

b 18 km

c 18 km/h

d 1 h 30 min

e 1 h 40 min

f A: $m = -\dfrac{1}{5}$

B: $m = 0$

C: $m = -\dfrac{2}{5}$

g Tama biked for half an hour at 12 km/h and covered 6 km. He then stopped for 50 minutes when he was 12 km from home. He cycled the final 12 km in half an hour and at a speed of 36 km/h.

4 a

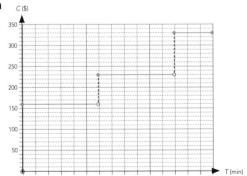

b $0 \le T$ (min) < 120: $C = 160$

$120 \le T$ (min) < 240: $C = 230$

$240 \le T$ (min) ≤ 300: $C = 330$

c $160.00 **d** $330.00

e His charges cover just the cost of his time, but they don't cover parts or materials needed to complete the job. If parts or materials were needed, he would have to charge more.

Simultaneous equations (pp. 58–65)

Solving simultaneous equations (pp. 58–59)

1 $x = 7, y = 3$ **2** $x = –4, y = 30$

3 $x = 2, y = 7$ **4** $x = 2, y = –3$

5 $x = 3, y = 2$ **6** $x = 4, y = 1$

7 $x = \dfrac{4}{5}, y = \dfrac{28}{5}$ **8** $x = \dfrac{11}{2}, y = 4$

Solving simultaneous equations using graphs (pp. 60–63)

1 a $15 **b** $0.50

c

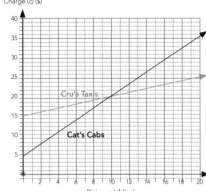

d Cru's Taxis: $c = 0.5d + 15$

Cat's Cabs: $c = 1.5d + 5$

e Cru's Taxis: $19

Cat's Cabs: $17

f Cru's Taxis: 16 km

Cat's Cabs: 12 km

g 10 km

$20

2 a

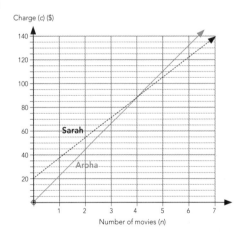

b $22

c Sarah: $c = 17n + 20$

Aroha: $c = 22n$

d Sarah pays $122 and Aroha pays $132. Aroha pays $10 more.

e Five movies.

To go to four movies costs both $88. When they go to five or more movies, the line for Sarah is below that for Aroha.

f Could reduce the annual fee to, say, $15, or reduce the cost per movie to, say, $16.

3 a

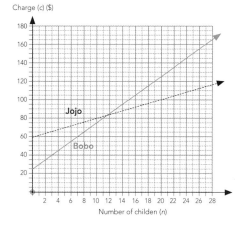

Charge (c) ($) vs Number of children (n)

b $25   **c** $5

d Bobo: $c = 5n + 25$
Jojo: $c = 2n + 60$

e $(11\frac{2}{3}, 88\frac{1}{3})$
Bobo is cheaper if there are fewer than 12 children at the party. Jojo charges less per child if there are 12 or more children at the party.

f Jojo could reduce his booking fee to, say, $50, or reduce his charge per child to, say, $1.50.

4 a

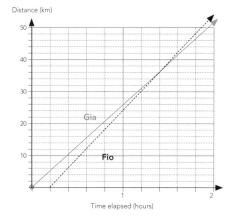

Distance (km) vs Time elapsed (hours)

b Gia: $d = 26t$
Flo: $d = 30t - 6$

c After 1.5 hours, the racers will both be 39 km from the start.

d Gia rides consistently at 26 km/h and leads for the first hour and a half until they have both cycled 39 km. At this point, Flo, riding at 30 km/h, overtakes her. Flo leads for the rest of the race and finishes after 1.87 hours (112 minutes). Gia takes 1.92 hours (115.38 minutes) to finish the race.

e We have assumed that both cyclists raced at constant speeds.

Forming and solving simultaneous equations (pp. 64–65)

1 Equations: $3c + 2a = 14.17$, $c + 5a = 15.99$
Solutions: $a = 2.60$, $c = 2.99$
So avocados were $2.60 and capsicums were $2.99.

2 Equations: $s + t = 85$, $t = 2s + 7$

Solutions: $s = 26$, $t = 59$
So Sam has $26 and Taylor has $59.

3 Equations: $V + L = 58$, $V = 3L + 2$
Solutions: $V = 44$, $L = 14$
So Victor is 44 and Lee is 14.

4 Equations: $a + c = 144$, $3c + 15a = 1524$
Solutions: $a = 91$, $c = 53$
So 91 adults and 53 children attended the fair.

5 Equations: $e = 3a$, $e = 2a + 3$
Solutions: $a = 3$, $e = 9$
So Miley has 3 Achieved and 9 Excellence results.

6 Equations: $R = 12L$, $\frac{R}{9} = L - 2$
Solutions: $R = 72$, $L = 6$
So 72 right-handed and 6 left-handed students were surveyed.

Linear inequalities (pp. 66–67)

1 $x > 9$
This means that x is **greater** than **9**.

2 $x \leq 12$
This means that x is **less** than or equal to **12**.

3 $x < 4$
This means that x is **less** than **4**.

4 $x \geq 3$
This means that x is **greater than or equal to 3**.

5 $x > -13$
This means that x is **greater than −13**.

6 $x \geq 2$
This means that x is **greater than or equal to 2**.

7 $x < -19$
This means that x is **less than −19**.

8 $x \leq \frac{5}{3}$
This means that x is **less than or equal to $\frac{5}{3}$**.

9 $x > -8$
This means that x is **less than −8**.

10 $x \leq 2$
This means that x is **less than or equal to 2**.

11 $2\frac{2}{3}x + 40 \leq 152$
$x \leq 42$
Maximum dimensions are 42 cm by 14 cm.

12 $x(18 + 27) \leq 200$
$x \leq 4\frac{4}{9}$
So 4 kōwhai trees and 14 hebes.

Rearrangement of expressions (pp. 68–69)

1 $r = \dfrac{P}{2\pi}$ **2** $d = vt$

3 $r = \sqrt{\dfrac{A}{\pi}}$ **4** $x = \dfrac{y - c}{m}$

5 $n = \dfrac{S}{180} + 2$ **6** $r = \sqrt{\dfrac{3V}{\pi h}}$

7 $b = \dfrac{2A - ha}{h}$ or $\dfrac{2A}{h} - a$

8 $C = \dfrac{5(F - 32)}{9}$ **9** $\cos A = \dfrac{a^2 - b^2 - c^2}{2bc}$

10 $r = \sqrt[3]{\dfrac{3V}{4\pi}}$

Quadratic expressions (pp. 70–86)

1	$x^2 + 8x + 12$	**2**	$x^2 + 6x - 7$
3	$x^2 + x - 12$	**4**	$x^2 - 14x + 45$
5	$x^2 + 12x + 36$	**6**	$x^2 + 11x + 24$
7	$x^2 - 10x + 25$	**8**	$x^2 - 17x + 72$
9	$x^2 + 8x + 16$	**10**	$x^2 - 49$
11	$x^2 + 22x + 121$	**12**	$x^2 - 36$
13	$x^2 + 16x + 64$	**14**	$-x^2 - 5x + 14$
15	$-x^2 + x + 12$	**16**	$-x^2 - 11x - 10$
17	$x^2 - 4x + 4$	**18**	$x^2 - 14x + 45$

Factorising quadratic expressions (pp. 72–73)

1	$(x + 2)(x + 7)$	**2**	$(x + 3)(x + 9)$
3	$(x + 4)(x + 10)$	**4**	$(x + 1)(x + 11)$
5	$(x + 10)(x + 12)$	**6**	$(x + 3)(x - 8)$
7	$(x + 11)(x - 4)$	**8**	$(x - 2)(x - 3)$
9	$(x - 8)(x - 10)$	**10**	$(x - 6)(x + 5)$
11	$(x - 5)(x + 5)$	**12**	$(x + 3)(x + 1)$
13	$(x + 13)(x + 2)$	**14**	$(x - 4)(x + 4)$
15	$(x - 2)(x + 2)$	**16**	$(x + 7)(x + 9)$
17	$(3 + x)(3 - x)$	**18**	$(x - 6)(x - 1)$

Solving quadratic equations (pp. 74–75)

1	$x = -6$ or 3	**2**	$x = 5$ or -2
3	$x = 1$ or 8	**4**	$x = -7$ or -4
5	$x = 10$ or -3	**6**	$x = 4$
7	$x = -7$	**8**	$x = 0.5$
9	$x = 4$ or -1	**10**	$x = 2$ or -8
11	$x = 0$ or 5	**12**	$x = 0$ or -9
13	$x = -5$ or -2	**14**	$x = 3$ or -1
15	$x = -8$ or 8	**16**	$x = -10$ or 8
17	$x = 17$ or $x = -4$	**18**	$x = 13$ or $x = 14$

Forming and solving quadratic equations (pp. 76–78)

1 $(x - 1)(x + 5) = 55$
$x = 6$ or 14
A length of –10 is impossible, so $x = 6$.
Base = 11 m. Height = 5 m.

2 $n^2 \times 7 = 448$
$n = 8$ or -8.
She could be thinking of 8 or –8.

3 $(c - 7)(c + 21) = 204$
$c = 13$ or -27
An age of –27 is not possible, so $c = 13$. Claudia is 13 years old and her dad is 41.

4 $\dfrac{3}{2}h^2 = 216$
$h = 12$ or -12
A height of –12 is impossible, so $h = 12$.
Base = 36 cm. Height = 12 cm.

5 $(w + 2)(w + 9) = 170$
$w = 8$ or -19
It is not possible to have –19 lollies, so $w = 8$.
Wally started with 8 lollies.

6 $x(x + 16) = 1380$
$x = -46$ or 30
A side of –46 is impossible, so $x = 30$.
Side = 46 mm.

7 $x^2 + (x + 7)^2 = (x + 8)^2$
$x = -3$ or 5
A side of –3 is impossible, so $x = 5$.
The sides are 5 m, 12 m and 13 m.

8 $x(x + 2) + 24 = (x + 2)(x + 4)$
$x = 4$
Gloria's age cannot be –4, so Gloria is 4 and her two siblings are 6 and 8.

9 $(8y \times 5y) - (3y \times 2y) = 1224$
$y = -6$ or 6
A length cannot be negative, so $y = 6$.
The big rectangle is 30 cm by 48 cm.

Plotting quadratic equations (pp. 79–81)

1

x	y
3	11
2	6
1	3
0	2
–1	3
–2	6
–3	11

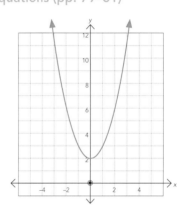

The vertex is a minimum/~~maximum~~ at (**0**, **2**).

2

x	y
4	11
3	6
2	3
1	2
0	3
–1	6
–2	11

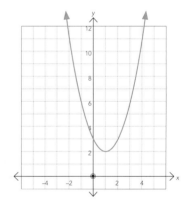

The vertex is a minimum/~~maximum~~ at (**1**, **2**).

3

x	y
1	–1
0	4
–1	7
–2	8
–3	7
–4	4
–5	–1

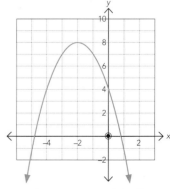

The vertex is a ~~minimum~~/maximum at (**–2**, **8**).

4

x	y
4	0
3	5
2	8
1	9
0	8
−1	5
−2	0

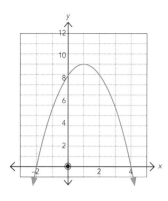

The vertex is a ~~minimum~~/maximum at (**1**, **9**).

5

x	y
4	12
2	3
0	0
−2	3
−4	12

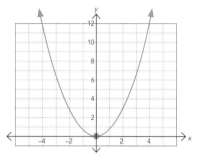

The vertex is a minimum/~~maximum~~ at (**0**, **0**).

6

x	y
3	11
2	5
1	3
0	5
−1	11

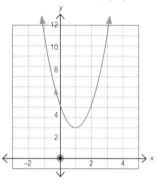

The vertex is a minimum/~~maximum~~ at (**1**, **3**).

Optimising by graphing quadratics (pp. 82–86)

1

x	Other side (120 − x)	Area: $y = \frac{1}{2}x(120 - x)$
10	110	550
20	100	1000
30	90	1350
40	80	1600
50	70	1750
60	60	1800
70	50	1750

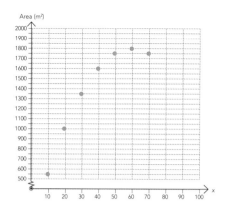

The sides that result in the maximum area are 60 m by 60 m. This will allow Sam to have an enclosure with an area of 1800 m².

2

Width (x)	Length (40 − 2x)	Area: y = x(40 − 2x)
6	28	168
7	26	182
8	24	192
9	22	198
10	20	200
11	18	198
12	16	192

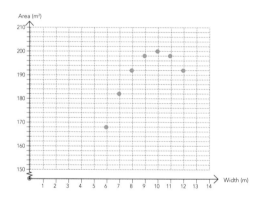

The maximum area Annie can enclose is 200 m², with a width of 10 m and a length of 20 m.

3

x	Height: y = 10x(9 − x)
1	80
2	140
3	180
4	200
5	200
6	180
7	140

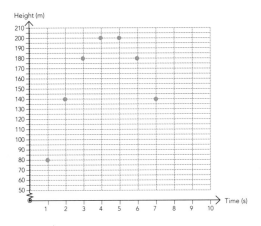

The rocket reached the maximum height of 202.5 m at 4.5 seconds.

Assumptions: There was no wind or rain. These would probably change the equation and reduce the height it reached.

4

x	Height: $y = 0.4(x - 2)^2 + 1$
0.5	1.9
1	1.4
1.5	1.1
2	1
2.5	1.1
3	1.4
3.5	1.9

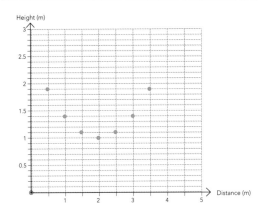

The lowest point is 1 m above the ground. The midpoint of the bar is at $x = 2$, so the bar must be 4 m long.

Geometry and space (pp. 87–117)
Fundamentals (pp. 87–90)
Words (p. 87)

Term	Description	Diagram
Acute angle	An angle that is less than 90°	
Obtuse angle	An angle that is greater than 90° but less than 180°	
Right angle	An angle that is exactly 90°	
Reflex angle	An angle that is greater than 180° and less than 360°	
Scalene triangle	A triangle with sides of three different lengths	
Isosceles triangle	A triangle with exactly two equal sides	
Equilateral triangle	A triangle with three equal sides	
Right-angled triangle	A triangle with an angle that is exactly 90°	
Supplementary angles	A pair of adjacent angles that add to 180°	
Complementary angles	A pair of adjacent angles that add to 90°	

Similar triangles (pp. 91–95)
Justification of similar triangles using angles (p. 92)

1 $\angle TVU = 50°$ $\angle SRQ = 49°$

These two triangles have equal angles so they are similar.

2 $\angle ACB = 68°$ $\angle FGH = 72°$

These two triangles have different angles so they are not similar.

3 $\angle ACB = 97°$ $\angle EDF = 40°$

These triangles have equal angles so they are similar.

4 $\angle FHG = 63°$ $\angle EDF = 53°$

These triangles have different angles so they are not similar.

Justification of similar triangles using sides (p. 93)

1 $\dfrac{ML}{OP} = \dfrac{25}{50} = \dfrac{1}{2}$

  $\dfrac{MK}{ON} = \dfrac{21}{42} = \dfrac{1}{2}$

  $\dfrac{KL}{NP} = \dfrac{18}{36} = \dfrac{1}{2}$

These triangles are similar because all their sides are proportional.

2

$$\frac{AC}{FH} = \frac{66}{130} = \frac{33}{65}$$

$$\frac{BC}{GH} = \frac{54}{108} = \frac{1}{2}$$

$$\frac{AB}{GF} = \frac{30}{60} = \frac{1}{2}$$

These triangles are not similar because not all of their sides are proportional.

3

$$\frac{BC}{FH} = \frac{10}{13}$$

$$\frac{AC}{FG} = \frac{8}{12.5} = \frac{16}{25}$$

$$\frac{AB}{GH} = \frac{6}{7.5} = \frac{4}{5}$$

These triangles are not similar because none of their sides are proportional.

4

$$\frac{BC}{DF} = \frac{90}{67.5} = \frac{4}{3}$$

$$\frac{AB}{EF} = \frac{74}{55.5} = \frac{4}{3}$$

$$\frac{AC}{DE} = \frac{72}{54} = \frac{4}{3}$$

These triangles are similar because all their sides are proportional.

Calculating lengths of unknown sides (pp. 94–95)

1
$$\frac{x}{22.5} = \frac{30}{45}$$
$$x = \left(\frac{30}{45}\right) \times 22.5$$
$$x = 15 \text{ mm}$$

2
$$\frac{y}{3} = \frac{4}{5}$$
$$y = \left(\frac{4}{5}\right) \times 3$$
$$y = 2.4 \text{ cm}$$

3
$$\frac{x}{2.4} = \frac{2.4}{1.8}$$
$$x = \left(\frac{2.4}{1.8}\right) \times 2.4$$
$$x = 3.2 \text{ cm}$$

4
$$\frac{AE}{96} = \frac{35}{56}$$
$$AE = 60 \text{ cm}$$
$$\frac{AB}{88} = \frac{35}{56}$$
$$AB = 55 \text{ cm}$$

5
$$\frac{h}{12.1} = \frac{1.65}{2.75}$$
$$h = \left(\frac{1.65}{2.75}\right) \times 12.1$$
$$h = 7.26 \text{ m}$$

I have assumed that the time between the two measurements was very short so the angle of the sun has not changed significantly. I have also assumed that the ground is level and horizontal.

6
$$\frac{x}{0.8} = \frac{1.5}{2}$$
$$x = \left(\frac{1.5}{2}\right) \times 0.8$$
$$x = 0.6 \text{ m}$$

The shorter post is 0.6 m, so the total timber length needed is 3.7 m.

I have assumed that there was no wasted wood and the wood lost in saw cuts was negligible.

Theorem of Pythagoras (pp. 96–102)

1	7.21 cm	**2**	29.93 mm
3	9.38 m	**4**	3.14 km
5	3.14 m or 313.85 cm	**6**	2.19 km or 21857.49 cm
7	6.06 cm	**8**	32.68 m

Multi-step and 2D practical problems (pp. 98–100)

1 BD = 48.734 m
CD = 50.99 m

2 AC = 4.996 cm
DC = 2.985 cm
AD = x = 2.01 cm

3 Other two sides: $\sqrt{882}$ and $\sqrt{512}$ cm.
$y = 37.34$ cm (2 dp)

4 AB + BC + CD + DA
$= \sqrt{45} + \sqrt{25} + \sqrt{52} + \sqrt{34}$
247.50 mm or 24.75 cm

5 $12 \times 1.532 = 18.28$ cm

6 BD = 13 cm   CD = 9.19 cm

7 $6.3 + 4.1 + 7.517 = 17.92$ km

8 $\sqrt{15^2 + 28^2} = 31.76$ m
The ladder needs to be 31.76 m long, so a 32 m ladder will reach the tenth floor.

9 $\sqrt{30.2^2 + 30.2^2} = 42.7$ cm
So the fabric is not quite square.

3D problems (pp. 101–102)

1 **a** 20.52 m   **b** 24.41 m
   c 28.65 m

2 AC = 4.751 km
AD = 5.37 km

3 CE = 0.5 x 35.355 = 17.678 cm
AE = 30.62 cm

4 19.05 cm   **5** 4.04 m

Trigonometry (pp. 103–111)

Finding sides (pp. 104–106)

1	16.71 cm	**2**	35.77 mm
3	4.29 km	**4**	1.83 m
5	41.31 mm	**6**	3.63 m
7	8.81 cm	**8**	84.44 cm
9	76.99 mm	**10**	72.19 km
11	31.65 m	**12**	1/4.02 mm

Mixing it up (p. 107)

1	99.30 cm	**2**	27.00 cm
3	10.72 cm	**4**	0.94 m
5	72.75 mm	**6**	3.77 cm
7	2.55 m	**8**	144.17 mm

Multi-step problems (pp. 108–109)

1 $\cos 67° = \frac{z}{2} \div 82$
$z = 64.08$ cm

2 $\sin 38° = \frac{5.7}{2} \div y$
$y = 4.63$ m

3 $AC = 16.095$ cm
$x = 13.18$ cm

4 $AC = 7.656$ m
$y = 6.20$ m

5 $AD = (\tan 50° \times 723) - (\tan 27° \times 723)$
$= 493.25$ m

6 $y = 11.01$ cm

7 $BC = 145.31$ mm
$\angle DBC = 39°$
$DC = 117.67$ mm

8 $y = \sin 50° \times 10$
$= 7.66$
$x = \cos 50° \times 10$
$= 6.43$
Coordinates: (6.43, 7.66)

Finding angles (pp. 110–111)

1 46.2°   **2** 50.9°
3 58.5°   **4** 39.9°
5 52.2°   **6** 28.9°
7 43.4°   **8** 77.5°

Putting it together (pp. 112–113)

1 30.58 cm   **2** 3.41 m
3 53.1°   **4** 692.55 mm
5 40.82 m   **6** 28.6°
7 313.18 mm   **8** 73.7°
9 $\angle ACD = 34°$
$x = 15.68$ cm
$BD = 8.77$ cm
$y = 5.91$ cm
10 $\angle ABE$ and $\angle CBD = 55.44°$
$y = 34.56°$
$BD = 10.33$ cm
$x = 5.5$ cm

Applications (pp. 114–115)

1 The anchor line $a = 50.22$ m (2 dp)
$50.22 \div 3 = 16.74$ m
So the length of anchor line is not long enough, as it's not three times the depth.
2 Height = $1.58 + 90 \times \tan 28°$
$= 49.43$ m
3 $\theta = 32.5°$
4 Height = $1.75 + 48.77$
$= 50.52$ m
5 $x = 4.4°$
This angle is less than 4.8°, so the ramp will comply with the building code.
6 Henare's eye height is 113.87 m above the water. The yacht is 168.83 m from the cliff.
7 $AD = 15.6538$ m
$BD = 8.677$ m (3 dp)

Bearings (pp. 116–117)

1 a $\angle ABC = 180° - 15° - 75°$
$= 90°$ (co-int $\angle$s add to 180°, // lines)
∴ Theorem of Pythagoras can be used to calculate the length of AC.
$CA = \sqrt{680^2 + 300^2}$
$= 743.23$ m (1 dp)

b $\angle BCA = \tan^{-1} \frac{680}{300} = 66°$ (0 dp)
∴ Bearing = $360° - (66° - 15°)$
$= 309°$

2 a $\angle ABC = 180° - 60° - 30°$
$= 90°$ (alt $\angle$s, // lines and $\angle$s on a line add to 180°)
∴ Theorem of Pythagoras can be used to calculate the length of AC.
$CA = \sqrt{3.37^2 + 7.41^2}$
$= 8.14$ NM (2 dp)

b $\angle BCA = \tan^{-1}\left(\frac{3.37}{7.41}\right)$
$= 24.5°$
Bearing = $360° - 120° - 24.5°$ ($\angle$s at a point add to 360°)
$= 215.5°$

c I have assumed that there are no currents or winds that will drive the yacht off course or alter the distance measurements.

Measurement (pp. 118–144)
Unit conversion (pp. 118–120)

Units of area (p. 119)

1 40 000 m²   **2** 2.64 m²
3 19 ha   **4** 68.30 cm²
5 1 970 000 mm²   **6** 9300 m²
7 0.0064 ha   **8** 3400 m²
9 145.35 m²   **10** 30 m

Units of volume (p. 120)

1 3 000 000 cm³   **2** 4260 mm³
3 0.0098 m³   **4** 5100 cm³
5 13.572 cm³   **6** 2370 cm³
7 0.019 L   **8** 7800 mm³
9 107 100 L   **10** 12 glasses

Two-dimensional shapes: revision (pp. 121–124)

Perimeter	Area
Perimeter = $2b + 2h$	Area = bh
Perimeter = $a + b + c$	Area = $\frac{1}{2}bh$
Perimeter = $2\pi r$ or πd	Area = πr^2
Perimeter = $2a + 2b$ or $2(a+b)$	Area = bh
Perimeter = $a + b + c + d$	Area = $\frac{b+d}{2} \times h$ or $\frac{h}{2}(b+d)$

1 Perimeter = 13 m
   Area = 7.74 m²

2 Perimeter = 12 cm
   Area = 5.32 cm²

3 Perimeter = 235 mm
   Area = 2961 mm²

4 Perimeter = 16.34 m (2 dp)
   Area = 21.24 m² (2 dp)

Compound shapes and shapes with holes (pp. 123–124)

1 Perimeter = 48 m
   Area = 108 m²

2 Perimeter = 228 mm
   Area = 1494 mm²

3 Perimeter = 3.54 m (2 dp)
   Area = 0.4886 m² (4 dp)

4 Area = 328 cm²

5 Perimeter = 12π = 37.70 cm (2 dp)
   Area = 75.40 cm² (2 dp)

Three-dimensional shapes: surface area and volume (pp. 125–144)

Shape	Volume	Surface area
Triangular prism	Volume = $\frac{1}{2}bh \times l$	Surface area = $2(\frac{1}{2}bh) + 3bl$
Cylinder	Volume = $\pi r^2 \times l$	Surface area = $2\pi r^2 + 2\pi r \times l$
Cone	Volume = $\frac{1}{3}(\pi r^2 \times h)$	Surface area = $\pi r l + \pi r^2$
Pyramid	Volume = $\frac{1}{3}(b^2 \times h)$	Surface area = $b^2 + 4(\frac{1}{2}b \times s)$
Sphere	Volume = $\frac{4}{3}\pi r^3$	Surface area = $4\pi r^2$

Surface area (pp. 126–128)

1 Net
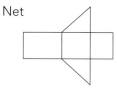
   1820 mm²

2

   2940.5 cm² (1 dp)

3

   0.2016 m²

4

   2000 cm²

5 111 036 mm²   6 3.382 m²
7 175 000 cm²   8 3770 cm²

Volume (pp. 129–132)

1 30 450 cm³   2 2494 cm³
3 1 595 000 mm³   4 0.2888 m³
5 1222 cm³   6 2.108 m³
7 4524 cm³   8 21.79 m³

9 198.7 m³   10 5481 cm³
11 1 040 000 mm³   12 0.03763 m³
13 2 065 000 cm³   14 1.642 cm³
15 0.3390 m³   16 38.7 cm³

Compound shapes: surface area and volume (pp. 133–135)

1 SA = 5.8² + 4(4.65 × 5.80) + 4($\frac{1}{2}$ × 5.80 × 6.3)
       = 214.6 cm²
   V = (5.80 × 4.65 × 5.8) + $\frac{5.8^2 \times 5.6}{3}$
     = 219.2 cm³

2 SA = 2(16 × 18) + 2(20 × 18) + 2(20 × 16) − (2 π × 4²)
          + (2 × π × 4 × 18)
       = 2288 cm²
   V = (16 × 18 × 20) − (π × 4² × 18)
     = 4855 cm³

3 SA = (2π × 0.62²) + (π × 0.62 × 1.15)
       = 4.655 m²
   V = ($\frac{2\pi \times 0.62^3}{3}$) + ($\frac{\pi \times 0.62^2 \times 0.97}{3}$)
     = 1.196 m³

4 SA = (π × 40²) + (2 × 40 × 40) + (π × 40 × 76)
       = 24 630 cm²
   V = (π × 40² × 40) + ($\frac{\pi \times 40^2 \times 65}{3}$)
     = 310 000 cm³

5 SA = (5 × 15²) + 4($\frac{1}{2}$ × 15 × 16.7)
       = 1626 cm²
   V = 15³ − $\frac{15^3}{3}$
     = 2250 cm³

Working backwards (pp. 136–137)

1 d = 8 cm   2 w = 28 cm
3 h = 10 cm   4 r = 0.5 m
5 h = 25 cm   6 d = 82 cm
7 w = 54 mm   8 h = 0.4 m

Estimation (pp. 138–141)

Your answers are likely to differ from these. If so, discuss them with your teacher.

1 Model: Sphere

   Formula: $V = \frac{4}{3}\pi r^3$

   Variable: r = 4.5 cm because the container looks slightly wider than it is tall, and the internal width will be less than the external width.

   Estimate of volume = $\frac{4}{3}\pi(4.5)^3$

   = 380 cm³ (2 sf)

   Assumption(s): That the container is spherical, but it isn't, so I might have underestimated its volume.

   Limitation(s): I have not accounted for the thickness of the plastic wall.

2 Model: Cone

   Formula: $V = \frac{1}{3}\pi r^2 h$

   Variables: r = 1.5 because the pile looks as though it is about the width of two shovel lengths.

   ISBN: 9780170477468

h = 1 m because the pile looks as though it is about two-thirds of a shovel length.

Estimate of volume = $\frac{1}{3}$ x π x (1.5)² x 1

= 2.4 m³ (2 sf)

Assumption(s): That the heap of soil is a perfect cone, which clearly it is not.

Limitation(s): I have not allowed for the air gaps between the soil clumps.

3 Model: Cylinder

Formula: $V = πr^2h$

Variables: r = 5 cm because the average diameter of the jug looks as though it is about a fifth of the height, and the inside diameter will be smaller than the outside diameter.

h = 20 cm because the milk looks as though it is about four-fifths of the total height.

Estimate of volume = π x 5² x 20

= 1600 cm³ (2 sf)

Assumption(s): That the jug is a cylinder, which it is not.

Limitation(s): This model applied only if the inside base is flat and horizontal.

4 Model: Cylinder

Formula: $A = πr^2h$

Variable: r = 8 cm because the length of the brush looks as though it is about the diameter of the circle.

h = 0.2 because paint is usually quite thick.

Estimate of volume = π x 8² x 0.2

= 40 cm³ (1 sf)

Assumption(s): That the puddle is circular, which it is not. I hope the volume outside a circular shape is the same as the gaps inside it.

Limitation(s): The model applied only if the paint is of a consistent thickness.

5 Model: Cone

Formula: $V = \frac{1}{3}πr^2h$

Variables: r = 3 cm because the external diameter of the rim is 8 cm, so the internal diameter looks as though it is about 6 cm.

h = 4 cm because it looks about half of the total diameter.

Estimate of volume = $\frac{1}{3}$ x π x 3² x 4

= 38 cm³ (2 sf)

Assumption(s): That the drink occupies an exact cone shape. The bottom of the glass is not cone shaped, so I have probably over-estimated the volume.

Limitation(s): The model applied only if the thickness of the glass is consistently 1 mm.

6 $V = b^2h$

= 7.3² x 18

= 959.2 cm³ (1 dp)

The carton is not an exact cuboid in shape because the sides bulge a little, which increases its internal volume. I have also assumed that

1000 cm³ of milk = 1 L. This is true for water, but might be slightly different for milk.

Applications (pp. 142–144)

1 Area = 2.35² – 4(1.1)²

= 0.6825 m²

2 a V = (1 x 0.6 x 0.45) – (0.9 x 0.5 x 0.40.09)

= 0.09 m³

b 10 L = 0.01 m³, so he will need 9 bags.

c SA = (2 x 1 x 0.45) + (2 x 0.6 x 0.45) +

(1 x 0.6 – 0.9 x 0.5)

= 1.59 m²

2 coats = 3.18 m²

d Volume paint = 0.2891 L (4 dp)

e V = 0.9 x 0.5 x 0.35 m³

= 0.1575 m³

= 157.5 L

3 a $πd + d$ = 100

d = 24.15 cm (2 dp)

b L = 2 x 24.14

= 48.28 cm

c Area remaining = (48.28 x 24.14) – 2 (π x ($\frac{24.14}{2}$)²)

= 250.1 cm²

4 a $V = \frac{1}{2}(\frac{4}{3}$ x π x 2.2³) – $\frac{1}{2}(\frac{4}{3}$ x π x 2³)

= 5.546 cm³ (3 dp)

b $V = (\frac{1}{3}$ x π x 1.75² x 8) – ($\frac{1}{3}$ x π x 1.6² x 7.8)

= 4.210 cm³ (3 dp)

Challenges (pp. 145–149)

1 $H = \sqrt{39.5}$ cm

Volume = $\frac{1}{3}$ x 7 x 7 x $\sqrt{39.5}$

= 102.65 cm³

Glass density = $\frac{2\,500\,000\ g}{1\,000\,000\ cm^3}$ = 2.5 g/cm³

Mass = 102.65 x 2.50

= 256.6 g (1 dp) or 0.2566 kg

Assumptions: I have assumed that the paperweight is exactly pyramidal. If, for instance, the tip is slightly rounded, or the base is not completely flat, then my model will not be accurate. I assume there were no bubbles in the glass. If there are, I will have overestimated the mass.

Limitations: This calculation of mass will be accurate only if the mass of a tonne of glass is exactly 2500 kg. It is possible that glass varies in composition, and if so the mass of the pyramid would be different.

Generalisation: Let b be the length of the base and h the vertical height.

Mass = volume of the square-based pyramid x density of glass

= $\frac{1}{3}b^2h$ x 2.5

= $0.8\dot{3}b^2h$

2 The first lap will take four times as long as the second lap.

First lap: 400 m takes $\dfrac{4 \times 320}{5} = 256$ s

$\therefore x = \dfrac{400}{256}$ m/s

Second lap: 400 m takes $\dfrac{320}{5} = 64$ s

Speed = $4 \times 1.5625 = 6.25$ m/s

He takes 64 seconds to run the second lap and his speed is 6.25 m/s.

Assumptions: He ran at a constant rate for each lap. This is unlikely because by the end of the second lap, he was probably tiring. At the end of the first lap, he instantaneously increased his speed from 1.5625 m/s to 6.25 m/s. In practice this would not happen because he would increase his speed over several seconds.

Limitations: These calculations would be accurate only if there were not changes in the direction or strength of the wind during his laps.

3 1 litre of water = 1000 cm³

500 litres of water = 500 000 cm³

Triangle depth = $\sqrt{70^2 - 35^2} = 60.62$ cm

Volume = $\dfrac{1}{2} \times 70 \times 60.62 \times L = 500\,000$

$L = 2.357$ m

Assumptions: That the trough was filled right to the top, which probably would not happen, so it may contain slightly less than 500 L. I have also assumed that the trough was shaped exactly as a prism.

Limitations: My model and calculation of the volume of water did not take into account the volume of any pipes or fittings that need to be fixed to the inside of the tank. If these are significant, they would reduce the volume of water in the tank.

Generalisation: $V = \dfrac{1}{2} \times t \times \sqrt{t^2 - \left(\dfrac{1}{2}\right)^2} \times L$

$= 1.2Lt\sqrt{\left(\dfrac{t}{2}\right)^2}$

4 $V = \dfrac{d}{t}$

$t = \dfrac{3.825 \times 10^5}{5.0 \times 10^4}$

$= 7.65$ hours

$= 7$ hours and 39 minutes

Assumptions: I have assumed that the moon was 3.825×10^5 km from Earth at the time, not further away or closer. If it was further away, I will have underestimated the time, but if it was closer, I will have overestimated the time.

Limitations: These calculations will be valid only if the rocket travels at an average speed of 5.0×10^4 km/hour.

Generalisation: In order to find the time (t) taken for an object to travel a distance (d) at a given speed (v), I used the formula $t = \dfrac{d}{v}$.

For finding the distance from Earth to the moon, this formula becomes $t = \dfrac{3.825 \times 10^5}{v}$.

5 Hemisphere: $\quad SA = 2\pi r^2$

$= 2 \times \pi \times 8^2$

$= 128\,\pi$

Cone: by Pythagoras, $L = \sqrt{12^2 + 8^2}$

$= 14.42$ cm

$SA = \pi rL$

$= \pi \times 8 \times 14.42$

$= 115.38$ cm²

Total SA = $243.38\,\pi$

$= 764.6$ cm²

$= 0.07646$ m²

Toy needs two coats, so SA to be covered for each toy = 0.1529 m²

Two litres of paint will cover 12 square metres.

Number of toys = $\dfrac{12}{0.1529} = 78.48$

Therefore 78 toys could be painted with two litres of paint.

Assumptions: I have based my calculations on a cone height of 12 cm, but the cone from which the tip was removed would have been taller. So I may have underestimated the surface area. So it's possible that there would be enough paint for only 77 toys.

Limitations: These calculations can be used only if one litre of paint covers 6 m². Paints vary in their coverage, so changing the paint might change the answer.

Practice tasks (pp. 150–160)

If your answers or reasoning are different, talk to your teacher.

Practice task one (pp. 150–152)
Number of nests

1 ha or 10 000 m² has 300 nests

$\therefore$ 8500 m² has $300 \times \dfrac{8500}{10\,000}$

$= 255$ nests

Surface area

$SA = 4\pi r^2$

$= 4\pi \times 2.25^2$

$= 63.62$ cm² (2 dp)

Amount of spray

$V = \dfrac{DS \times 10^{-2}}{3}$

$= \dfrac{255 \times 63.62 \times 10^{-2}}{3}$

$= 54.1$ L (1 dp)

This assumes that the whole of her 8500 m² is suitable for wasp nests. If some is not, this would reduce the estimated number of nests and the amount of spray needed.

ISBN: 9780170477468

I have assumed that the wasps are actually at a density of 200 nests per hectare.

I have assumed that wasp nests are exactly spherical.

Ladder length

$$\text{Length} = \frac{3.5}{\sin 75°}$$

$$= 3.63 \text{ m (minimum length} \Rightarrow \text{need to round up)}$$

I have assumed that the ground between the ladder and the wall is flat and horizontal.

Amount charged for 5 hours

Amount charged if Percy does the job and Beatrice pays promptly

$$= \$368 \div 1.15 = \$320$$

Amount to pay if Ernie does the job:

Equation: $C = 55x + 60$

Cost = $55 \times 5 + 60 = \$335$

∴ provided Beatrice pays promptly, Percy will be $15 cheaper.

New method for calculating Percy's charge.

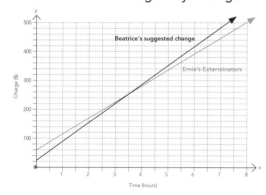

Point of intersection

Ernie's Exterminators: $y = 55x + 60$

Percy's Pest Services: $y = 65x + 25$

$$65x + 25 = 55x + 60$$
$$10x = 35$$
$$x = 3.5$$
$$\therefore y = \$252.50$$

This means that for 3.5 hours (3 hours and 30 minutes), they both charge $252.50.

For less than 3.5 hours (3 hours and 30 minutes), Ernie is cheaper, but for more than this interval, Percy is cheaper.

Practice task two (pp. 153–155)

Maximum number of people

Area of tent = $\pi \times 1.45^2 = 6.61 \text{ m}^2$ (2 dp)

$$\text{Maximum number of people} = \frac{6.61}{1.4} = 4.72 \text{ (2 dp)}$$

∴ the maximum number of people is 4.

I have assumed that the area calculations are for adults. If children are sleeping in the tent, they would need less space, and you could probably fit 5 or 6.

Surface area

Slant height $(l) = \sqrt{1.6^2 + 1.9^2}$

$$= 2.48 \text{ m (2 dp)}$$

$$SA = \pi r l$$
$$= \pi \times 1.6 \times 2.48$$
$$= 12.5 \text{ m}^2 \text{ (1 dp)}$$

Spray needed and composition

Volume of spray needed = $12.5 \times 50 \text{ mL}$

$$= 125 \text{ mL}$$

3:2 means there are 5 parts

$$\text{Each part} = \frac{625}{5} = 125 \text{ mL}$$

∴ he will need 375 mL of waterproofing liquid and 250 mL of water.

I have assumed that the top of the tent is a perfect cone, which is unlikely. If there is any sag in the roof, you would need slightly more spray. I have also assumed that there is no wastage of spray.

Peg distance

Angle of elevation = $\tan^{-1} \left(\frac{0.3}{1.2}\right) = 14.0°$

So the angle of elevation will be more than the recommended 13°.

The recommended angle is probably conservative and accounts for all weather conditions. If the tent is in a sheltered position and the weather is forecast to be calm, it probably wouldn't matter if the angle is steeper than that recommended. To comply, Hemi would need to place his pegs further away from his tent.

$$\text{Distance} = \frac{0.3}{\tan 13°} = 1.299 \text{ m}$$

So in practice his pegs would need to be 1.3 m away from the tent.

Inflatable mattresses

Better Bargains: Price = $\$120 \times 85\% = \$102.00$

$$\text{Winnie's Warehouse} = \frac{120}{1.15} = \$104.35$$

I have assumed that the mattresses are the same brand and that the conditions on the sale at the two outlets are the same.

Practice task three (pp. 156–158)

Scale

2 cm on plan = 2 m on the house

2 cm on plan = 200 cm on the house

∴ Scale = 1:100

Kitchen/dining room is 5.5 cm × 100 = 5.5 m

Brace length

$$HC = \sqrt{1.8^2 + 2.8^2}$$
$$= 3.329 \text{ m (3 dp)}$$
$$EC = \sqrt{3.329^2 + 2.5^2}$$
$$= 4.163 \text{ m}$$

So a minimum of 4.17 m timber will be needed. This assumes that there is no wastage. In practice they should probably buy a 4.2 m length.

Number of sheets

$$\text{Number of sheets} = \frac{0.2}{4.0 \times 10^{-4}}$$
$$= 500 \text{ sheets}$$

This assumes that the sheets fit together exactly and there are no air gaps between them. It is likely that there are slightly fewer than 500 in the stack.

Paint volume and cheapest way

$$\text{Paint needed} = \frac{24 \text{ m}^2}{10 \text{ m}^2} = 2.4 \text{ L}$$

2 L + 1 L costs $149.98.
2 L + 2 L costs $179.98.
4 L costs $142.00
∴ it is cheapest to buy a 4 L tin.

Dimensions of vegetable garden

Width (x)	Length (12 – 2x)	Area: y = x(12 – 2x)
1	10	10
2	8	16
3	6	18
4	4	16
5	6	10

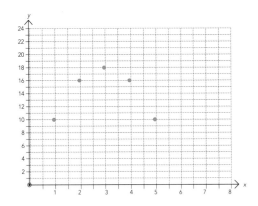

Maximum occurs at the point (3, 18).
This means that the short sides of the rectangle are 3 m which means the length is 6 m, and the total area is 18 m².

Practice task four (pp. 159–160)

Cost of yarn

Ratio yarn length:finished cord length = 5:3
∴ Length of yarn needed for each cord = $12 \times 21 \times \frac{5}{3}$
= 420 cm = 4.2 m
∴ 80 pairs of poi need 160 × 4.20 m = 672 m

A cone contains 600 m of yarn so she will need 2 cones.
$$\text{Cost} = 2 \times \frac{41.40}{1.15} = \$72$$

Area of fabric

Surface area of each poi = $4\pi r^2$
$$= 4\pi \times 4^2$$
$$= 201.1 \text{ cm}^2$$

Circle of fabric must have an area of 201.1 cm²
Area of this circle = $\pi \times r^2$ = 201.1 cm²
∴ Radius of fabric circle needed to cover a poi =

$$\sqrt{\frac{201.1}{\pi}} = 8 \text{ cm}$$

She needs a 2 cm margin, so the radius of each circle must be 10 cm.
Each poi will need a square of fabric which is 20 cm x 20 cm = 400 cm².
Total area = 160 x 400 cm² = 64 000 cm²
$$= 6.4 \text{ m}^2$$

This assumes that there will be no waste of fabric and that the shape can be cut into 20 x 20 cm squares.

Number of cushions needed

Volume of 160 poi = $160 \times \frac{4}{3} \times \pi \times 4^3$
$$= 42\,893 \text{ cm}^3$$
Volume of 1 cushion = $0.018 \times 100^3 \text{ cm}^3$
$$\text{Number needed} = \frac{42\,893}{18\,000}$$
$$= 2.4$$
∴ 3 cushions are needed.

Number of adults and children

Let a be the number of adults and c be the number of children.

$a + c = 2263$	①
$30a + 18c = 58\,854$	②

From ① $a = 2263 - c$
Substitute for a in ②
$30(2263 - c) + 18c = 58\,854$
$67\,890 - 30c + 18c = 58\,854$
$12c = 9036$
$c = 753$
$a + 753 = 2263$
$a = 1510$
So there were 1510 adults and 753 children.